Process Validation for Medical Devices

A Pocket Book

Emmet Tobin © 2017

ISBN-13:978-1977834010

ISBN-10:1977834019

Contents

CHAPTER 1
VALIDATION PLANNING

Introduction 17
Generic Benefits 18
Types of Validation Plans 18
Matrix or Family Approaches 20
Changes to the Validated State 20
Changes Impacting Operational Qualification – OQ 21
Changes Impacting Performance Qualification – PQ 22
Validation Plans and FMEA 24
Qualification Plan 25

CHAPTER 2
FACILITIES AND UTILITIES QUALIFICATION

Introduction 26
Risk an Impact Assessment 27
Qualification Levels 29
Critical Utilities 30
GMP Zoning 36
Types of Contamination 37
Room Air Classifications 38
Compliance Tests for GMP Zones 39

CHAPTER 3

EQUIPMENT AND SOFTWARE VALIDATION

Introduction 43
The Goal of Equipment Qualification 43
Equipment Classification 46
Risk/Impact Assessment 46
Installation Qualification 47
Operational Qualification 47
The Validation Lifecycle 47
Types of Validation 48
Requirements Specifications 49
Factory Acceptance Testing (FAT) 51
Equipment Qualification (EQ) Protocols 51
Post Execution Review 52
Software Validation 54
Software Validation and GAMP 54
Key Terms 55
Computer System Validation Life Cycle 56
Regulatory Review 60
Specification Hierarchy 61
System Categorisation 64
GAMP Considerations 66
Risk Assessments 66
Quality Risk Matrix 67
Traceability Matrix 68
General Requirements 68
Uninterrupted Power Supply 70
Types of UPS 71
Software Source Code Review 72
Calibration 73
Deviations 73
Requalification 74

CHAPTER 4
PROCESS VALIDATION

Introduction 75
Test Methods and Process Validation 77
Stages of Process Validation 78
Fundamentals of Process Validation 79
Process Validation and Dominance Factors 80
Process Operational Qualification (OQ-P) 82
Protocol Approval Check list 83
Process Performance Qualification 84
Yield Data (aka Process Yield Data) 84
Continued Process Verification 85
Revalidation (or Maintaining a Validated State) 85
Other Scenarios – Maintaining a Validation State 86
Acceptance Criteria 86
Validation Strategies 87
Principles of Worst Case Selection 87
Requalification 87
Process Validation - Examples of Deficiencies 88
Case Studies on Process Validation 89
Case Study - CNC Grinder–Performance Qualification (PQ) 89
Case Study – Cleanline (OQ-P/PQ) 91

CHAPTER 5
PACKAGING VALIDATION

General Requirements for Design and Development 96
Inputs 98
Outputs 99
Sterile Packaging 100
Stage 2 – Material(s), Equipment and Process technology
Supplier Requirements 101
Materials 102
Equipment and Process Technology 103
Stage 3 - Material Performance and Suitability Testing 104
Test Methods 104

Stage 4 - Stability Testing 108
Stage 5 - Packaging Performance Testing 109
Stage 6- Packaging Validation 110
Medical Packaging Process Validation 110
Statistical Methods 112
General Principles of Blister Packaging Validation 112
Variable Print Packaging 113
What is Variable Print? 113
What Is Fixed Print? 113
Packaging Definitions 114
Further Reading 114

CHAPTER 6
TEST METHOD VALIDATION

Introduction 116
Examples of Test Method Validations 116
What is test method validation? 117
Why should TMV be performed? 117
When should methods be validated? 118
Regulatory Overview 120
US Food and Drug Administration 120
W.H.O 127
ISO 13485 129
Definitions and Key Concepts 138
New Test Methods 140
Changes to Existing Methods 140
Accuracy 143
Precision 144
Ruggedness 144
Representative/Continuous Sampling 145
Range 146
Resolution 146
Probability Of False Alarms P (Fa) 146
Probability Of Misses P (M) 147
Validation Protocols 147
What Can Impact the Accuracy of a Test Method? 147

General MSA requirements 149
Variable MSA Studies 150
Attribute MSA Studies 151
Measurement Capability Index 152
Template Appendix I 155
Template Appendix II 161

CHAPTER 7
MEASUREMENT

Introduction 168
Care & Maintenance of Instruments 189
Manual Measuring Instruments 170
Slip Gauges 172
The Sine Bar 172
Other Applications of the Sine Principle 173
Vision Measuring Systems 173
Coordinate Measuring Machines 173
Optical Comparators 174
Digital Comparators 175
Geometric Tolerancing 175

CHAPTER 8
ISO 13485

Introduction 179
Standard Clauses 187
CE Marking 201
Device Classification 206
Summary Of Rules 207

CHAPTER 9
LEAN BASICS

Introduction 218
A Roadmap to Lean 219
The History of Lean 220
Understanding Flow 223
The Goal of Lean? 224
Lean & Six Sigma 225
Toyota Production System (TPS) 225
Value Stream Mapping 226
Poka Yoke 227
Customer Driven Companies 231
Jigs and Fixtures 234
Common Poka-Yoke Tools 234
Kaizen 236

Published by Solo Validation Resources Limited, © Copyright 2017, First Edition

This book is for general guidance purposes only. It is the responsibility of individuals, companies and organisations to implement the necessary legal and regulatory requirements relevant to their industry.

Solo Validation Resources Limited will take no responsibility for the application or interpretation of content contained herein.

Notes

Notes

Notes

CHAPTER 1
VALIDATION PLANNING

Introduction

Validation planning plays a key role in the qualification and validation of new equipment and processes. It also has a role in established processes and is used to plan and manage the ongoing validation requirements within a company. So why the need for validation plans? Firstly, the requirement for validation within medical device and pharmaceutical companies is a legal and regulatory one. The Food and Drug Administration (FDA) stipulates validation as a regulatory requirement of Good Manufacturing Practices (GMP) for both pharmaceuticals (21 CFR 211) and medical devices (21 CFR 820). Validation plans act like a qualification plan that can be used to document strategies, technical rationales and key deliverables. They are a regulatory requirement for medicinal products manufactured in the United States and Europe.

Although not stated in 21 CFR Part 820 (Medical Devices), validation planning is an important activity that helps to document the validation strategy and is commonplace with medical device manufacturers. In Europe, EudraLex (V4 GMP) is the collection of rules and regulations governing medicinal products in the European Union which also require validation planning.

All equipment, processes, facilities and utilities that are GxP impacting need to be qualified. To facilitate the validation efforts, a Validation Plan (VP) creates a roadmap and structure to meet the validation requirements. For simple processes or simple equipment qualifications, a stand-alone validation plan may not be required and can be captured within a protocol or change control. The requirements of validation plans can be driven by a procedure which may be local to a site or factory, or may be corporate and applicable to multiple sites. Consistency of requirements can also be managed by the use of an approved validation plan template.

Generic Benefits

Apart from regulatory or procedural requirements to create a validation plan, there are many other beneficial reasons to complete one. A validation plan acts as a top level document that can pull together the many references, protocols, reports and rationales that make up a project. It is also a powerful asset to introduce new staff and team members to a project or process who need to get up to speed quickly and comprehensively. Validation plans are often the first documents an auditor will request to see in relation to a new process or new product introduction. They force the various stakeholders to sit down and agree upon the strategy and any technical rationales required to deliver a successful project.

Types of Validation Plans

Validation plans can be divided into three different types or configurations. Depending on the validation activity or the project, a validation plan may take the form of a (1) Site Validation Plan (aka Site Master Validation Plan) (2) Master Validation Plans (MVPs) or (3) Individual Validation Plans (VPs).

From the outset, it is important to highlight that different companies may adopt different terminology with regard to validation planning. Typically, large companies will have a site validation plan aka site master validation plan or equivalent document.

A site MVP details the products, processes and associated validation protocols and reports for a manufacturing site/factory. It is the overarching validation plan. Typical components of a site MVP include: description of products and processes, test methods (analytical, physical), specifications, an up-to-date list of utility qualifications, equipment qualifications and process validations.

An MVP encompasses all aspects of a validation strategy and may include multiple processes, multiple pieces of equipment/machines that require validation. MVPs are common for new product introductions. Although not stated in 21 CFR Part 820, MVPs are useful in documenting the validation status.

An individual validation plan generally details the validation strategy of one piece of equipment or machine, therefore, an individual validation plan tends to be limited to a handful of pages. Nonetheless, it is valuable in documenting the validation approach and is a central document that can detail process development reports, specifications and so on.

Matrix or Family Approaches

A family or matrix approach to validation can be used where similar products are produced using the same equipment and processes. A particular product size or product configuration may be selected to represent the "worst-case" product. Therefore, by qualifying the worst case, all of the other products within the family are considered validated. Matrix or family approaches must be clearly documented with technical rationale provided in advance of any qualification activities. This can be addressed in a validation plan or within a protocol. Alternatively, a technical report or product development report can be created and referenced. Taking a family of products, the worst case product might be, the smallest, the largest, the heaviest, or the product requiring the greatest precision and so on.

Changes to the Validated State

Revalidation may be necessary under the following conditions:

- change(s) in the actual process that may affect quality or its validation status
- change(s) in the product design which affects the process
- transfer of processes from one facility to another
- change of the application of the process
- change in materials
- change in a manufacturing agent (cleaning agent, oils, greases, coolant, detergents etc.)

The need for revalidation should be evaluated and documented. Evaluation needs to consider historical results from quality indicators, product changes, process changes, changes in external requirements (regulations or standards) and other such circumstances, as applicable. Revalidation may not be as extensive as the initial validation if the situation does not require that all aspects of the original validation be repeated.

Changes Impacting Operational Qualification - OQ

For changes made to a qualified process, it is necessary to evaluate whether worst case conditions exist. The VOSA may be used to facilitate this evaluation. For changes impacting worst-case conditions, conduct an OQ study in order to challenge those outputs and related inputs at worst-case conditions. If no worst-case conditions exist for the affected outputs, then an OQ of the change is not required. Rationale for not conducting an OQ must be documented in the PQ report.

Narrowing or tightening process parameters within a qualified range does not typically require qualification. Take a scenario that seeks to widen process parameters outside of the qualified range. Suggested minimum validation activity would include:

➢ Both operational and performance qualification runs to qualify the new range.
➢ All product outputs impacted by the change should be tested and challenged as would happen in an initial validation.

Changes Impacting Performance Qualification (PQ)

If changes are proposed to a qualified process, the impact of the changes should be assessed and documented. Changes to critical process parameters may require a full re-qualification depending on the level of change that is proposed. Supporting studies may be required to support the proposed change such as engineering studies or testing.

<u>What Factors Should Determine the Content of a Validation Plan?</u>

An initial barrier to developing an effective validation plan is the availability of information. When a project is in the concept stage, decisions may still be required on the direction of the project, e.g. what technology will be used? How many new machines will be purchased? What is the timeline?

At the project level, it is likely that a project charter or some scoping document is available in draft copy or is being developed. These documents can help create a validation plan and give it an initial framework. It might also suggest minimum requirements for particular disciplines. An analytical VP would differ to a VP for a new manufacturing facility. In addition, a formal or approved template will prompt the author to include the right information and content.

<u>Content of a Site Validation Master Plan</u>

A site VMP tends to be quite top level in nature with mostly generic and non-specific content. It is not used to describe what the strategy involves for a new project or modification to equipment or process. Some suggested headings include:

Products: a list of products and a brief description of them. Some background on their uses.

Equipment and Processes: a description of the different equipment, technologies and processes used in the manufacture of the different products.

Facilities/Utilities: a brief description of the facilities and utilities onsite.

Overview of Process Validation: a general explanation of the validation policy e.g. risk-based etc.

Summary of Validation Reports for Site: a listing or summary of the completed validations.

Content of an Individual Validation Plan

Depending on the type and scope of validation, the following headings should be considered for inclusion (as applicable):

Products: as above.

Equipment and Processes: as above.

Facilities/Utilities: as above.

Overview of Process Validation: as above.

Validation Inputs: inputs include URSs, design history files, project charters etc.

Validation Deliverables: typically, a list of both activities and documents to be approved/completed can be included here.

Test methods and Specifications: a list of relevant test methods used during process testing or product testing.

Process Development: if process development studies are to be completed, this section can be populated. Studies can be referenced here with rationale for the validation strategy outlined.

Risk Management: if risk assessments are completed for the equipment or process, the critical items can be listed here along with the risk references.

FAT (Factory Acceptance Testing): this section can document the need for a FAT or not e.g. "No FAT is required for this equipment/process as it is an off-the-shelf item" OR "The FAT will be completed prior to shipping".

SAT (Site Acceptance Testing): this section can document scope and approach to the SAT.

Validation Plans and FMEAs

A Process FMEA (pFMEA) involves assessing each of the process steps, documenting the things that can go wrong at each process step and determining potential consequences for the product. In turn, this information is used to adopt the appropriate risk mitigation and control measures. These mitigations may be applied up front (redesign of a jig or fixture), during the process (e.g. monitoring and controlling a critical process parameter), or after the process at inspection.

Qualification Plan

A Qualification Plan (QP) describes all the qualification measures and at which stage of the qualification the verification will be completed. A qualification plan typically contains detailed descriptions of the necessary test measures and a description of the interdependencies of the individual tests. References to other test documents such as FAT or SAT and a description of the deviation management may also be integrated into the qualification plan. In some instances, there may not be a need or a requirement for a qualification plan. A validation plan can also serve to detail the qualification strategy. A simple table or matrix can be used to map out the requirements and qualification activities.

Test	FAT	COMISSIONING	SAT	IQ/OQ
Drawings review	X			
Documentation	X			X
Calibration		X		X
Walk down	X		X	
Alarm testing	X			X
Functional testing	X		X	X
Utilties		X	X	

Figure: Simple qualification plan

The FDA provides clear definitions on the four types of validation which include:

(1) Prospective validation
(2) Concurrent validation
(3) Retrospective validation
(4) Revalidation

A validation plan should identify what type of validation is to be conducted, e.g. prospective or concurrent.

CHAPTER 2

FACILITIES AND UTILITIES QUALIFICATION

Introduction

Facilities and utilities qualifications are typically pre-requisites to the validation of manufacturing equipment and systems. Much of the activity that deals with establishing a facility or building that is fit for purpose is managed under the broad heading of commissioning and qualification (C&Q). The terms C&Q are often used interchangeably and in practice some overlap in activity is expected.

Commissioning can be defined as the planned, documented, and managed engineering approach to the start-up and handover of facilities, systems, and equipment to the end-user. It must deliver a safe and functional environment that meets the pre-defined design and user requirements.

In strict terms, qualification is more concerned with the confirmation and documentation showing that equipment or systems are properly installed and functional. Qualification forms part of validation, but the individual qualification steps do not equal a validated process. The establishment of a user requirements specification (URS) and detailed design specifications ensure that the building or facility will meet end users' needs and that it is fit for the intended purpose. It also provides a level of protection to the contracting company responsible for the project or facility construction. Post-URS approval requires an approved Design Qualification (DQ). This provides verification and a documented record that the proposed design is suitable for the intended purpose. Further verification including IQ/OP/PQ should be applied as required based on the system impact and criticality of facilities/utilities.

Risk an Impact Assessment

A risk-based qualification process should assess the potential of a system to impact the product quality. The boundaries of any system (HVAC, compressed air supply etc.) should be identified in order to help establish the scope of any system and determine if it has a direct, indirect or no impact on product quality.

Direct Impact: a system that can directly impact product quality.

Indirect Impact: where a system is not expected to directly impact the product quality but supports or is ancillary to a direct impact system.

No Impact: a system that does not directly impact product quality and does not support a direct impact system.

Example of HVAC System Boundaries

For a HVAC System supplying a classified area, only once the air enters the room must the air quality meet the classified designation. The Critical Quality Attributes (CQAs) are routinely monitored through the Environmental Monitoring Program and the Critical Process Parameters (CPPs) should be monitored through the calibrated and validated Environmental Monitoring System (QBMS). The Direct Impact (level 1) for the HVAC systems are indicated on the boundary diagram shown below.

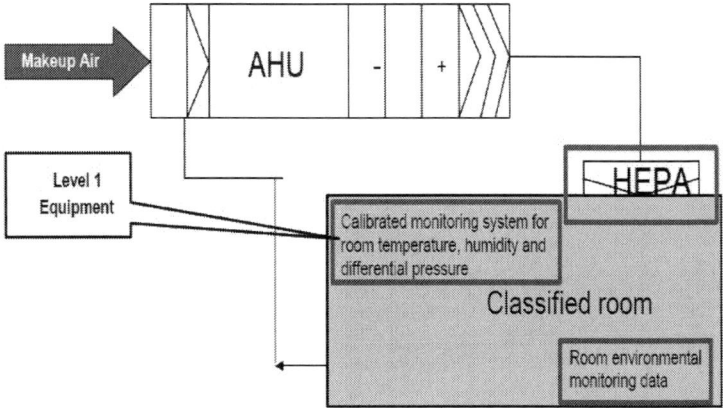

Figure: HVAC System Boundary Diagram (Level 1). Each individual system is represented by a green box. Separate qualifications should be performed for each one. The room environmental monitoring system is typically called a Building Management System (BMS). The calibrated monitoring system for room temperature, humidity and differential pressure is called a QBMS. Where the "Q" stands for quality indicating the system is used to monitor critical parameters.

Qualification Levels

Qualification levels are often used within companies to classify the criticality of equipment or systems. Level 1 requires the highest level of verification.

Level 1: a system where an undetected change in system performance poses a significant risk to the product and product safety. Level 1 systems require the highest degree of qualification and validation. This should include URS/DQ/IQ/ OQ/ PQ.

Level 2: a system where a change that may be detected in system performance poses a significant risk to product and product safety. These systems require a level of qualification including IQ, however OQ and PQ testing may not be required. This should be based on the intended use of the system, impact on product quality and overall risk.

Level 3: All other systems.

Typically IQ or equivalent testing is sufficient. Note: other requirements or qualifications should be based on risk.
The level of qualification and validation testing required for any system should be based on a risk assessment, examining the criticality of the system and environment. Risk assessments should consider the following points:

- Building design and construction features
- System boundaries and complexity
- Potential product impact
- Environmental controls and monitoring systems
- Potential impact to operator safety
- Type of qualification/validation (e.g. prospective, concurrent, or retrospective)

Controlled-not-classified (CNC) environments, utilities, and facility control systems also require adequate qualification/validation. Again, the impact on product quality should be determined in order to shape any validation. Routine monitoring test locations as well as alert and action levels should be determined in advance of any validation for environmental monitoring or utility systems.

Critical Utilities

The process of identifying critical utilities can be done with the application of direct impact, in-direct impact and no impact definitions (see previous section "risk and impact assessment"). Risk assessments, CQAs and CPPs should also help identify critical utilities. When critical utilities are required as part of manufacturing and processing, the following points should be examined during the requirements and design stage:

- Materials of construction
- Internal surface finishes
- System sizing
- Flow rates, dead legs, drainage etc.

Compressed Air

Compressed air is used for valve actuation, instrument air, process air, and clean air to name but a few applications. Only the point-of-use filtration and the gas quality instrumentation should be classified as level 1. When flow or pressure is a CPP, the measurement/monitoring should be performed by the system into which the gas is flowing. Additionally, the CQAs and CPPs should be routinely monitored through the calibrated monitoring system. For compressed air, the potential CPPs are listed below. For the physical system being evaluated, the use and the application of the compressed air will determine which (if not all) CPPs are needed to ensure the system produces product of the desired quality.

    Hydrocarbons
    Moisture
    Particulates
    Temperature

Deionised Water

CPPs typical for a deionised water system include:

- Pressure
- pH
- Conductivity
- Level
- TOC
- Flow
- Temperature
- Resistivity

HVAC

Heating, ventilation and air-conditioning (HVAC) plays an important role in ensuring the manufacture of quality products. Furthermore, HVAC systems also provide comfortable conditions for operators based in the manufacturing environment. HVAC system design influences the layout of airlock positions and doorways. In turn, airlocks, entrances and exits have an effect on room pressure differential cascades and cross-contamination control. The prevention of contamination and cross-contamination is an essential design consideration of the HVAC system. In view of these critical aspects, the design of the HVAC system should be considered at the concept design stage of a manufacturing plant.

Temperature, relative humidity (RH) and ventilation should not adversely affect the quality of products during their manufacture and storage, or the proper functioning of equipment.

CPPs for HVAC systems include:

    Temperature
    Humidity
    Particle count – viable and non-viable
    HEPA filter certification/leak test/air flow rates
    Room differential pressures

Key Definitions

Alert limit: a value reached when the normal operating range of a critical parameter has been exceeded, indicating that corrective measures may need to be taken to prevent the action limit being reached.

At-rest: a condition where the installation is complete with equipment installed and operating in a manner agreed upon by the customer and supplier, but with no personnel present.

Cleanroom: an area (or room or zone) with defined environmental control of particulate and microbial contamination, constructed and used in such a way as to reduce the introduction, generation and retention of contaminants within the area.

Containment: a process or device to contain product, dust or contaminants in one zone, preventing it from escaping to another zone.

Contamination: the undesired introduction of impurities of a chemical or microbial nature, or of foreign matter, into or onto a starting material or intermediate, during production, sampling, packaging or repackaging, storage or transport.

Point extraction: air extraction to remove dust with the extraction point located as close as possible to the source of the dust.

Pressure cascade: a process whereby air flows from one area, which is maintained at a higher pressure, to another area at a lower pressure.

Relative humidity: the ratio of the actual water vapour pressure of the air to the saturated water vapour pressure of the air at the same temperature expressed as a percentage. More simply put, it is the ratio of the mass of moisture in the air, relative to the mass at 100% moisture saturation, at a given temperature.

Turbulent flow: turbulent flow, or non-unidirectional airflow, is air distribution that is introduced into the controlled space and then mixes with room air by means of induction.

The Displacement Concept (low pressure differential, high airflow)

This concept is commonly found in production processes where large amounts of dust are generated. Under this concept the air should be supplied to the corridor, flow through the doorway, and be extracted from the back of the cubicle. Normally the cubicle door should be closed and the air should enter the cubicle through a door grille, although the concept can be applied to an opening without a door. The velocity should be high enough to prevent turbulence within the doorway resulting in dust escaping. This displacement airflow should be calculated as the product of the door area and the velocity, which generally results in relatively large air quantities.

Note: This method of containment is not the preferred method, as the measurement and monitoring of airflow velocities in doorways is difficult.

Pressure Differential Concept (high pressure differential, low airflow)

The pressure differential concept may normally be used in zones where little or no dust is being generated. It may be used alone or in combination with other containment control such as a double door airlock. The high pressure differential between the clean and less clean zones should be generated by leakage through the gaps of the closed doors to the cubicle. The pressure differential should be of sufficient magnitude to ensure containment and prevention of flow reversal, but should not be so high as to create turbulence problems.

In considering room pressure differentials, transient variations, such as machine extract systems, should be taken into consideration. A pressure differential of 15 Pa is often used for achieving containment between two adjacent zones, but pressure differentials of between 5 Pa and 20 Pa may be acceptable. Where the design pressure differential is too low and tolerances are at opposite extremities, a flow reversal can take place. For example, where a control tolerance of ± 3 Pa is specified, the implications of rooms being operated at the upper and lower tolerances should be evaluated. It is important to select pressures and tolerances such that a flow reversal is unlikely to occur. The pressure differential between adjacent rooms could be considered a critical parameter, depending on the outcome of risk analysis.

The limits for the pressure differential between adjacent areas should be such that there is no risk of overlap in the acceptable operating range, e.g. 5 Pa to 15 Pa in one room and 15 Pa to 30 Pa in an adjacent room, resulting in the failure of the pressure cascade, where the first room is at the maximum pressure limit and the second room is at its minimum pressure limit. Low pressure differentials may be acceptable when airlocks (pressure sinks or pressure bubbles) are used to segregate areas.

The pressure control and monitoring devices used should be calibrated and qualified. Compliance with specifications should be regularly verified and the results recorded. Pressure control devices should be linked to an alarm system set according to the levels determined by a risk analysis. Manual control systems, where used, should be set up during commissioning, with set points marked, and should not change unless other system conditions change. Airlocks can be important components in setting up and maintaining pressure cascade systems and also to limit cross-contamination. Airlocks with different pressure cascade regimes include the cascade airlock, sink airlock and bubble airlock

GMP Zoning

Selecting a suitable classification for a room or manufacturing facility depends on several factors. Firstly, it can be said that sterile products require a more stringent set of criteria than non-sterile products. However, there is an extensive range of products and medical devices that are sterile but are used in different ways and consist of different materials and technology. Some sterile products are single use only and used for short term purposes and then disposed of. Other sterile products are used subcutaneously for longer periods or even require implantation. Therefore, the design of a facility along with its HVAC specification must be appropriate to the product being manufactured. High risk products require greater control. The goal of facilities and HVAC systems is to minimise contamination and the associated risks. Using a "sterile versus non-sterile" rule of thumb is not adequate when classifying a room or facility. Standards including EN ISO 14644-1 and guidelines such as EU cGMP Guidelines EudraLex volume 4 Annex 1 (2008) should be consulted in order to fully understand the requirements of each ISO classification and grade of room.

ISO classifications do not specify room occupancy states but when a designation is applied, the occupancy state must be stated in the relevant documentation or procedure. The most relevant European Guideline (Annex 1 of the EU cGMP Guideline) lists four classification grades and their associated particulate limits in the 'at rest' and 'in operation' conditions. In general, for the sterile and non-sterile products, similar classes are applied, but in non-sterile production the producer could assign their classes, having similar particulate concentration, temperature, pressure etc. but lower air-change rate could be used.

Types of Contamination

- cross contamination (of a product/material with another product/material)
- non-microbial particulate contamination (non-viable particles)
- biological/microbiological contamination (viable particles/micro-organisms)

Factors Influencing Contamination Cleanliness Levels in the Manufacturing Processes:

- process
- air cleanliness
- personnel hygiene and clothing
- work practices
- material design (material of construction, surface finishes, room finishes, equipment, open system/enclosed system
- utensils, etc.)
- material cleanliness

Room Air Classifications

Room Air Classification (By Limits of Particulate Contamination)

ISO CLASS	FDA	cCMP	Permissible particle number in 1 m3					
			0,1 µm	0,2 µm	0,3 µm	0,5 µm	1 µm	5 µm
1			10	2				
2			100	24	10	4		
3	1		1,000	237	102	35	8	
4	10		10,000	2,370	1,020	352	83	
5	100	A	100,000	23,700	10,200	3,520	832	29
6	1,000	B	1,000,000	237,000	102,000	35,200	8,320	293
7	10,000	C				352,000	83,200	2,930
8	100,000	D				3,520,000	832,000	29,300
9						35,200,000	8,320,000	293,000

Figure: Table showing ISO classes and EudraLex Grades A-D. Note: The EU guidance given for the maximum permitted number of particles in the "at rest" corresponds approximately to the ISO classifications.

	Maximum permitted number of particles per m³ equal to or greater than the tabulated size			
	At rest		In operation	
Grade	0.5 µm	5.0 µm	0.5 µm	5.0 µm
A	3 520	20	3 520	20
B	3 520	29	352 000	2 900
C	352 000	2 900	3 520 000	29 000
D	3 520 000	29 000	Not defined	Not defined

Figure: maximum permitted airborne particle concentration for each grade. Showing both "at rest" and "in operation" conditions. (EU V4 Annex 1)

Room Air Classification (By Limits of Microbial Contamination)

Grade	Recommended limits for microbial contamination (a)			
	air sample cfu/m^3	settle plates (diameter 90 mm) cfu/4 hours (b)	contact plates (diameter 55 mm) cfu/plate	glove print 5 fingers cfu/glove
A	< 1	< 1	< 1	< 1
B	10	5	5	5
C	100	50	25	-
D	200	100	50	-

The HVAC systems help maintain the viable (microbial) limits within a specific area. These limits are defined in Annex 1 of the EU GMP Guide as shown below.

Compliance Tests for GMP Zones

Test	Requirements
Particle count test	Test covers verification of cleanliness. Dust particle counts to be carried out and result printed. The no. of readings and positions of tests should be defined in accordance with ISO 14644-1 Annex B5
Air pressure difference	This test is used to verify non cross-contamination. Log of pressure differential readings to be produced or critical plants should be logged daily, preferably continuously. A 15 Pa pressure differential between different zones is recommended. Refer to ISO 14644-3 Annex B5

Airflow volume	To verify air change rates. Airflow readings for supply air and return air grilles to be measured and air change rates to be calculated. Refer to ISO 14644-3 Annex B13
Airflow velocity	To verify unidirectional flow or containment conditions. Air velocities for containment systems and unidirectional flow protection systems to be measured. Refer to ISO 14644-3 Annex B4
Filter leakage tests	To verify filter integrity) Filter penetration tests to be carried out by a competent person to demonstrate filter media, filter seal and filter frame integrity. Only required on HEPA filters. Refer to ISO 14644-3 Annex B6
Containment leakage	To verify absence of cross-contamination) Demonstrate that contaminant is maintained within a room by means of: • airflow direction smoke tests • room air pressures. Refer to ISO 14644-3 Annex B4
Recovery	To verify clean-up time) Test to establish time that a cleanroom takes to recover from a contaminated

	condition to the specified cleanroom condition. Should not take more than 15 min. Refer to ISO 14644-3 Annex B13
Airflow visualization	To verify required airflow patterns. Tests to demonstrate air flows: • from clean to dirty areas • do not cause cross-contamination • uniformly from unidirectional airflow units Demonstrated by actual or video-taped smoke tests. Refer to ISO 14644-3 Annex B7

Further reading

> ➤ ISPE Baseline Guide: Commissioning and Qualification, (2001)
> ➤ ISO 14644-1: International Organisation For Standardisation – Cleanrooms and Associated Controlled Environments. Part 1: Classification of Air Cleanliness
> ➤ ISO 14644-3: International Organisation For Standardisation – Cleanrooms and Associated Controlled Environments. Part 3: Test Methods
> ➤ ISO 14644-4: International Organisation For Standardisation Cleanrooms and Associated Controlled Environments: Part 4: Design, Construction and Start-Up

- EudraLex, Vol 4, Annex 1: EU Guide to Good Manufacturing Practice (EU GGMP) Governing Medicinal Products for Human and Veterinary Use, Annex 1 – Manufacture of Sterile Medicinal Products
- EudraLex, Vol 4, Annex 10: Manufacture Of Pressurised Metered Dose Aerosol Preparations For Inhalations
- ISPE Good Practice Guide: ISPE Good Practice Guide, HVAC (2009)
- EN1822:2009: European Standard For HEPA Filter Classification

CHAPTER 3

EQUIPMENT AND SOFTWARE VALIDATION

Introduction

The principles outlined in this chapter can be applied to manufacturing equipment and systems, test equipment and lab equipment. Equipment used in any manufacturing, testing or activity that has the potential to impact product quality needs to be properly commissioned and installed to a full validated state. The definition of the word "validation" should not be underestimated as it is the key driver of why processes and systems must be validated within regulated industries. It is evident as one progresses through the validation life-cycle that the definition and its key terms provide the framework to achieving a validated state. So here it is:

Validation is "Establishing documented evidence that provides a high degree of assurance that a specific process will consistently produce a product meeting its pre-determined specifications and quality attributes".

The Goal of Equipment Qualification

The ultimate goal of Equipment Qualification is to ensure that equipment is fit for its intended use. Therefore, equipment is validated to confirm it functions as intended and meets all requirements to manufacture product safely and consistently. FDA requires that "Each manufacturer shall ensure that all equipment used in the manufacturing process meets specified requirements and is appropriately designed, constructed, placed and installed to facilitate maintenance, adjustment, cleaning and use". In other words all manufacturing equipment, support facilities, measuring and test equipment must be "qualified". (FDA 21 CFR 820.70 (G))

Equipment qualification protocols are developed to document this testing and hence provide evidence on the functionality and consistency of the equipment. There are two distinct parts within the scope of equipment qualification, installation qualification and operational qualification. Often these subparts are abbreviated to IQ and OQ. Other combinations such as IOQE and IQ/OQ can be encountered within industry. This is often defined in a company's procedure or SOP relating to equipment validation.

A User Requirements Specification (URS) is often used to document the "specified requirements" of a particular piece of equipment. A URS can then be used as an input document when equipment qualification is required. While a URS document can be extensive covering areas such as equipment functionality, utility requirements, safety features, software specs etc. not all requirements documented in a URS will need to be verified or validated. Critical requirements should be identified early and should always be verified.

In short, equipment qualification is confirmation via documented evidence that the particular requirements for a specific intended use can be consistently fulfilled under anticipated conditions.

Often referred to as the three "Cs" of validation; confirmation, consistency and conditions (anticipated) are key themes that validation must address. Confirmation is addressed by the process of completing a formal validation. When it's done, it is documented and available for review to auditors. To assess consistency, there must be a number of batches or "runs".

44

Typically, there are minor batch-to-batch differences or variations between batches. These differences can be as a result of setup or raw material differences. Process validation must ensure that despite minor changes, there is consistency between batches, with product meeting specifications. Controlled or anticipated conditions are the machine or process settings that are known, documented and controlled during the manufacture of products.

The conditions during manufacture should stable and in control. Later on in this book we will see that settings must remain within the process window or validated window.

What Is Equipment Installation and Operational Qualification IQ/OQ?

Establishing documented evidence that all key aspects of the process equipment installation adhere to the manufacturer's approved specifications and any recommendations of the supplier of the equipment are suitably considered.

The process/equipment operates as intended and all user requirements are adequately fulfilled. When creating a URS it is important to:

- Specify hardware requirements
- Describe software requirements
- Determine the scope of operating parameters
- Identify safety requirements
- List calibration requirements
- Itemise materials of construction (if required)
- Stipulate the sequence of operation as required

URS documents should be reviewed by appropriately experienced personnel familiar with the product, process and equipment. This ensures critical functions and operations will be identified.

Equipment Classification

Custom: Custom equipment is where the design and requirements are specific to a customer's requirements. Often they are complex as "off-the-shelf" equipment does not meet the customer's needs. With customisation, the potential risk, or potential for manufacturing errors to occur increases.

Off-the-Shelf: Equipment which is available from suppliers that is ready made and does not require customisation.

Risk/Impact Assessment

Prior to any equipment qualification activities a risk or impact assessment should be completed. This is to identify all risk factors which may impact the product quality, performance of the process and ultimately the application of the product (with or by a patient). A risk-based approach to equipment qualification is best practice. The most commonly used process is Failure Mode Effect Analysis Methodology (PFMEA). It is also useful to categorise the potential risks into customer impact and process impact.

Installation Qualification

We have previously defined Equipment Qualification (EQ) and the two components to it (IQ and OQ) as "Establishing by documented evidence that all key aspects of the equipment installation meets the manufacturer's specification"). IQ is required in order to ensure that the equipment is installed, positioned and sited in a manner that is safe and in-line with manufacturer's recommendations. Once a piece of equipment is sited, it must be integrated into the utilities that are required to operate the equipment. An example of typical checks is listed on the following page.

Operational Qualification

The second element of equipment qualification now must be considered; equipment-operational qualification. This is "Establishing by documented evidence that the equipment operates per specifications and over the required ranges and to required tolerances". Equipment is also tested to ensure alarms and controls operate as required and intended. Some typical checks included in an equipment-operational qualification are testing of alarms, control system testing, utility failures and functional and operational testing.

The Validation Lifecycle

The Validation lifecycle refers to the requirement to control and document all validation activities from conception and URS stage to the retirement of equipment or a process. The lifecycle approach ensures compliance throughout the life of the process/equipment while maintaining a validated state throughout the application of change control.

With most equipment, systems and processes it is best practice to complete all qualification and validation activities in advance of the manufacture of any products for sale, commercial use and use in certain trials.

Types of Validation

The FDA provides clear definitions on the four types of validation which are explained below.

Prospective Validation

Establishing documented evidence in advance of process implementation that a process or system operates as intended. This is the preferred approach and is most common when new products must be validated before commercial manufacturing.

Concurrent Validation

Establishing documented evidence that a processes operates as intended, based on information generated during process implementation. Concurrent means that the outputs and performance of the system are monitored at the time of manufacturing which can include commercial lots.

Retrospective Validation

Retrospective validation is used for facilities or processes that have not completed formal validation. Historical data or a retrospective review can provide the evidence that the process or facility is operated as intended. This type of validation is uncommon.

Revalidation

Revalidation involves the re-execution of validation activities in order to maintain a validated state. This can be a result of substantial changes to product attributes, specification or changes to the manufacturing process itself. Other reasons a partial or full revalidation may be required involve instances where product quality issues have increased.

Requirements Specifications

While quality system regulations state that design input requirements must be documented, and specified requirements must be verified, the regulations do not further clarify the distinction between the terms "requirement" and "specification." The URS is the starting point for any piece of equipment or machine. The URS drives the project to conclusion and should take into account all of the factors relating to the product and process.

A requirement can be any need or expectation for a system or for its software. Requirements reflect the needs of the customer, and may be: (1) market-based, (2) contractual, or (3) statutory, as well as a (4) company's internal requirements. There can be many different kinds of requirements (e.g., design, functional, implementation, interface, performance, or physical requirements).

For example, take the scenario where a company wishes to purchase a thermal oven. From a statutory perspective, the company will ensure that the equipment is CE marked and is built in accordance with all EU directives as this is a fundamental requirement for equipment used within the EU. Similarly, validation is a statutory requirement for all medical devices and pharmaceuticals. Thus, it will be the responsibility of the customer to validate the equipment and the process. Internal requirements may include the need for suppliers to be certified to a QMS, or that they operate as a limited company.

Software requirements are typically stated in functional terms and are defined, refined, and updated during the development phase. Success in accurately and completely documenting software requirements is a crucial factor in successful validation of the resulting software. A specification is defined as "a document that states requirements." It may refer to or include engineering drawings or other relevant documents *21 CFR 820.3(Y).

There are different kinds of written specifications:

- User requirements specifications
- System requirements specification
- Software requirements specification
- Software design specification
- Software test specification
- Functional design specification

All of these documents establish "specified requirements" and are design outputs for which various forms of verification or validation are required. The URS must also define non-software requirements and hardware. Non-functional requirements such as maintainability and usability can also be included. There should be a clear distinction between mandatory regulatory requirements and optional features. The URS should be understood and agreed by both the user and supplier.

Suggested Prerequisites to Equipment Qualification

Prior to formal equipment installation and operational qualification (IQOQ), there are a number of engineering activities that can be completed. Although work is required up front in order to complete these activities such as preparing engineering test protocols, they will benefit the qualification stage. The completion of some or all of the above activities will help identify issues prior to formal equipment qualification. Essentially, an FAT protocol is like an early draft of an IQOQ-Equipment Protocol.

Factory Acceptance Testing (FAT)

An FAT or Factory Acceptance Test is an engineering activity. The purpose of the FAT is to verify the equipment or system meets the requirements of the URS. From the validation engineers perspective, it can be a learning activity and an opportunity to gather data, documentation and supporting design documents that will prove valuable during the equipment and process validation of the equipment. SAT is an engineering activity that is completed at the site of the vendor or equipment manufacturer, post FAT.

Equipment Qualification (EQ) Protocols

Protocol Preparation

Thorough and careful preparation is critical in order to successfully complete a qualification without deviations. In the preparation of the protocol, the URS is a key document. Often quotations, design documents, vendor drawings and owner manuals can contribute to the test and verifications to be completed during EQ.

Protocol Approval

Approval is always required prior to executing an equipment qualification protocol. Approvers should be aware that they are signing for the accuracy and content of the whole document. It is strongly advised that prior to final approval and execution of a protocol, a dry-run or trial is completed to ensure the test methods and acceptance criteria are accurate.

Post Execution Review

Upon execution of a protocol, timely review is advised in order to catch any errors or omissions. The person reviewing the protocol should not be the same person who performed the test. This review is best completed by a quality engineer; however, each organisation should identify personnel responsible for EQ reviews.

- Some points to remember when completing protocols:
- Ensure the protocol is fully approved prior to execution
- Ensure personnel are trained on the protocol (if required) and trained to the specific work instructions
- Ensure all team members, contractors etc. sign the signature log
- Observe safety precautions and wear PPE as required
- Ensure other employees that may be impacted by qualifications are aware that a qualification is in progress
- Ensure that any test product required in support of the qualification is identified, segregated and stored to internal standards
- Check that accurate work instructions are available (some companies may allow redlined copies to be used)
- Complete all tests in the protocol
- Always use indelible ink

- Carefully check each result against any acceptance criteria
- Ensure all test equipment is validated within calibration prior to use
- Handwritten comments should be signed and dated per GDP
- Deviations should be written up at the time of observation
- Data records and attachments should be identified with the protocol number, signed, paginated and dated
- When data is transcribed it should be verified by a second person. The source of the data should also be recorded
- All product manufactured should have relevant batch documentation as per normal production conditions

Equipment Qualification Reports

On completion, equipment installation and operational protocol reports are required. The format of any report largely depends on company specific procedures. If the protocol is an executable document (results are hand written in) then the executed version can be deemed the report. A summary report may be required but this depends on the requirements within the company or organisation.

The typical requirements of a completed EQ validation include:

- Equipment qualification protocol
- Equipment qualification protocol (executed)
- Raw data
- Attachments (examples of attachments include: material certs, calibration certs, CE Certs and MSDS

Software Validation

Where there is potential to affect product conformance to requirements or where software or IT systems provide support to aspects of quality management, validation is required. Most companies categorise software validations to account for the different applications of software and IT systems. For example, enterprise systems, such as the drawing package SolidWorks, would be validated in a different manner to manufacturing systems that contain software (a.k.a. embedded software).

"Embedded" software is where the software is integrated into the manufacturing equipment. Embedded software is typically validated during the equipment qualification stage, process validation stage or test method validation. Enterprise software falls outside of equipment or process validation but does require validation if it impacts product quality or is used to make quality decisions. Standalone systems such as ERP (Enterprise Resource Planning) systems also require validation.

Software Validation and GAMP

Good Automated Manufacturing Practice (GAMP) is a set of guidelines for manufacturers and users of automated systems in regulated industries. These guidelines are particularly important for the medical device, pharmaceutical and biopharmaceutical industries. The application of GAMP and validation of automated systems in manufacturing helps ensure that regulated medical devices and medicinal products have the required quality and are manufactured according to good practices, meet regulatory and legal requirements and ensure patient safety. GAMP ensures quality is in-built into each stage of the manufacturing process. Therefore, GAMP has a place in all aspects of automation and production, including the handling of raw materials, control of facilities and equipment etc.

Key Terms

Automated System: Term used to cover a broad range of systems, including automated manufacturing equipment, control systems, automated laboratory systems, manufacturing execution systems and computers running laboratory or manufacturing database systems. The automated system consists of the hardware, software and network components, together with the controlled functions and associated documentation. Automated systems are sometimes referred to as computerised systems.

Commercial Off-the-Shelf (COTS): Configurable programs and stock programs that can be configured to specific user applications by "filling in the blanks", without (COTS) altering the basic program.

Computer System Validation: A process that confirms by examination and provision of objective evidence that the computer system conforms to user needs and intended uses. System validation is a process for achieving and maintaining compliance with GxP regulations and fitness for intended use by adoption of life cycle activities, deliverables, and controls.

GAMP 5: Is a set of guidelines that offers a risk-based approach to ensuring the compliance of GxP-impacting computerised systems.

V- Model: Is a development process which sets out a roadmap of stages and deliverables during a project.
21 CFR Part 820: FDA requirements pertaining to medical devices.

User Requirement Specification, URS: The URS is a critical document that defines the requirements of the computerised system and agreement to the requirements.

Software Requirement Specification, SRS: An SRS can be written to interpret the requirements of a URS and how they relate to the requirement or how the requirement is met in practical terms regarding software.

Functional Design Specification, FDS: A functional design specification is a document that specifies how particular requirements are met – this can be a combination of how the equipment/process operates mechanically/automatically etc. An FDS is typically written in response to a URS.

Computer System Validation Life Cycle

The Computer System Validation Life Cycle refers to all activities from initial concept to retirement of a computer system. The life cycle of the system includes the defining of, and performance of activities in a systematic way from conception, requirements, development or configuration, testing, release and operational use.

The four GAMP life cycle phases include:

Concept
Planning and project stage
Operation
Retirement

The concept stage is concerned with understanding the need or the problem to be addressed. We will see that the User Requirement Specification (along with other specifications) and the initial risk assessment help to drive a project forward in a systematic manner. The most common life cycle approach for computerised and automated systems is the V-Model. The GAMP based V-model lays out a roadmap which facilitates the validation of equipment and automated systems.

The planning and project stage involves the planning of the validation effort required to implement the system into the business area(s) based on identification and approval of system concept. This phase includes assessments of the regulatory and system risks, supplier assessment, development of validation strategies, identification of deliverables that will be generated, definition of the business process the system will support as well as the user requirements which the system will fulfil.

Design, development and configuration of the hardware and software are also required to meet the system requirements as per specifications. In case of custom software components, this effort could also include detailed software design and developmental testing to ensure readiness for verification testing.

The verification stage confirms that specifications have been met and releases the system for use. This phase will involve multiple stages of reviews and testing depending on the system type, the development method applied and its use. Once verification activities have begun any changes to the system must be captured through change control.

On successful completion of the verification activities, the system is then released for effective use. The test strategy and other verification activities will vary widely between simple equipment and more complex customised/configurable systems. The verification and validation approach is typically agreed and detailed in the validation planning stage. The VP can be updated accordingly as the project develops with more detail being added. Alternatively, a test strategy document or matrix could be written to provide more specific test plans.

Verification deliverables vary based on the complexity and level, or customisation of the system in question. Corporate or company specific procedures also shape the required activities to be completed and reported. Some generic deliverables are listed below:

- Approval, execution and review of test protocols
- Writing and approving SOPs for operation and maintenance of the system
- Traceability matrix
- Completion of any risk mitigations (e.g. updates to FMEA etc.)
- Validation summary report(s)

Validation reporting requirements vary depending upon the scope of the system and should also be driven by a procedure and template. The validation plan can also outline the deliverables and what needs to be addressed in the report. A Validation Summary Report (VSR) must be written which summarises the results of executing the VP. The documents created for the validation activities summarise (or point to summaries) of the testing performed. Finally, the VSR indicates the acceptance of the system/equipment by the user and by the project team stating that the equipment is released for commercial operation/production.

The operation phase supports the need to maintain compliance and fitness for intended use after the system is released for normal use. It is important to ensure the system remains within a continued validated state. All proposed or necessary changes to the system must be assessed and controlled as part of a change control process. Once the system has been accepted and released for use, the operation phase begins. This phase consists of maintaining the system's compliant state and fitness for intended use through the control of the procedures supporting the system's operational use.

During the operation phase the below activities are typically completed:

- Ongoing training
- Preventative maintenance
- Service management and performance monitoring
- Change control
- Periodic review
- Maintaining system security
- Records management
- Calibration

The retirement phase involves the planning and proper management of activities relating to the removal of systems from service (shutdown). The retirement should take into account the storage of any data and any data migration that needs to occur prior to retirement. The retirement plan, if needed, will outline the retirement strategy from the roles and activities that will be conducted to the removal of the system for use. A Retirement Summary Report is produced that documents the results of the activities defined in the retirement plan including:

- Retirement Plan and timelines
- Summaries of any data migration activities

> Identification of the storage location of documentation relating to the system
> Obsoleting of SOPs

It must be stressed that GAMP is a set of principles, a set of guidelines that aims to achieve compliant computerised systems that are fit for intended use. GAMP Guidelines differ to 21 CFR QSR regulations as they are not legal or statutory requirements. However, they represent industry best practice and compliment the validation efforts that are legal requirements and statutory requirements.

Regulatory Review

Software validation is a requirement of the quality system regulation, 21 Code of Federal Regulations (CFR) Part 820. Validation requirements apply to:
(1) software used as components in medical devices
(2) software that is itself a medical device
(3) software used in production of the device or in implementation of the device manufacturer's quality system.

Note: EU GMP Annex 11, provides information on the inspection of 'computerised systems'.
In addition, computer systems used to create, modify, and maintain electronic records and to manage electronic signatures are also subject to the validation requirements. Such computer systems must be validated to ensure accuracy, reliability, consistent intended performance and the ability to discern invalid or altered records. The regulated user should be able to demonstrate through the validation evidence that they have a high level of confidence in the integrity of both the processes executed within the controlling computer system and in those processes controlled by the computer system within the prescribed operating environment.

60

Specification Hierarchy

An equipment URS can define the requirements of a computerised or automated system along with the operation, function, process and safety mandated by the customer. If the system is bespoke and complex, an SRS may be written to more clearly detail the software requirements, automation and functionality. Similarly, an FDS (functional design specification) can be written to address how mechanical or physical processing occurs.

URS-to-SRS

Scenario: a URS is written to specify the requirements for an automated blister packaging line in a medical device company. The URS details the following:

URS R1.0 – The machine shall be capable of operating in various modes to allow the manufacture of product and other debugging activities.

In turn, an SRS can be written to interpret the URS, for example, the SRS shall have the following modes:

SRS 1.1 Run empty mode - in this mode the equipment does not accept any new product.

SRS 1.2 Production mode – every station operates within the machine.

SRS 1.3 Bypass mode - where any operation can be disabled. Another two examples of a URS requirement being transposed to an SRS requirement are shown below:

Example 1: URS Requirement

URS R1.0: In the event of E-Stop activation, all sequencers shall be maintained and shall retain the sequence step that they were in at the time of E-Stop activation.

SRS Requirement

SRS 1.0 The lot count and lot integrity must be maintained after E-stop activation.

Example 2: URS Requirement

URS R1.0 The system shall use password protection.
SRS Requirement

SRS 1.1 Basic machine functionality (cycle start/stop, fault reset, manual operations) require no security.

SRS 1.2 As required, user IDs will be assigned to security groups for authentication. Authentication will be via active directory authentication against B+L domain accounts.

SRS 1.3 An auto-logoff feature shall be incorporated in the design.

Examples of Security Requirements:

- Three levels of access required, operator, and engineer and maintenance.
- Engineer - access to all screens, to modify process settings.
- Maintenance - access to functions required to perform machine maintenance activities.

- Operator - restricted access, does not have access to change process settings.
- Different access levels will require different passwords.
- No security will be required for basic operations (start/stop).
- A user auto-logoff feature shall be incorporated in the design. The auto-logoff time shall be configurable.
- A soft copy of program settings must be provided with delivery of the equipment.

Examples of HMI Requirements:

- The reject count and yield must be displayed on the HMI screen.
- Real-time readings for all critical parameters shall be visible on the HMI Screen.
- All critical parameters shall be adjustable via the HMI Screen.
- The status of each door should be visible on the HMI screen.

Examples of EHS Requirements:

- Activated E-Stops shall be clearly displayed on the HMI screen with a suitable alarm message generated.
- Activated E-Stops shall result in no further movement of the system until the E-stop is reset and all alarms are cleared.

System Categorisation

GAMP 5 makes provision for four categories of software in order to distinguish the level of customisation/configurability that exists across software serving different functions.
GAMP Software Category 1, Operating systems
GAMP Software Category 2, Non-configured software
GAMP Software Category 4, Configurable software packages
GAMP Software Category 5, Custom software

GAMP Software Category 1, Operating Systems

Category 1, operating systems, covers established, commercially available operating systems.

These are not subject to validation themselves; the name and version of the operating system must, however, be documented and verified during Installation Qualification (IQ). Application software hosted on operating systems need to be validated.
GAMP Software Category 2, Non-configured Software
Category 3 covers commercially available, standard software packages and "off-the-shelf" solutions for certain processes. The configuration of the software packages should be limited to adaptation to the runtime environment (for example network and printer connections) and the configuration of the process parameters. The name and version of the standard software package should be documented and verified in an Installation Qualification (IQ). Special user requirements, such as security, alarms, messages, or algorithms must be documented and verified in an Operational Qualification (OQ).

GAMP Software Category 4, Configurable Software Packages

GAMP Software Category 4, Configurable Software Packages Category 4 covers configurable software packages that allow special business and manufacturing processes. This involves configuring predefined software modules. These software packages should only be considered as belonging to Category 4 if they are well-known and mature. Normally, a supplier audit is necessary. If this is not available, the software packages should be handled as Category 5. The name, version, and configuration should be documented and verified in an Installation Qualification (IQ). The functions of the software packages should be verified in terms of the user requirements in an Operational Qualification (OQ). The validation plan should take into account the life cycle model and an assessment of suppliers and software packages.

GAMP Software Category 5, Custom Software

GAMP Software Category 5, Custom Software Custom/Bespoke Software (GAMP Software Cat 5) is software that contains custom code designed or modified specifically for a particular customer. As the code is custom it presents a greater risk. This risk must be mitigated with the right approach to the validation.

GAMP Considerations

Correctly assigning a GAMP software category to equipment, a system or process is an important activity that should be completed early on in the planning stage of a project. There must be some degree of familiarity with the equipment or system. The manufacturer or vendor can be a source of information that may help the designation. In many cases, companies create tools or processes that help determine what GAMP software category applies. These have different names such as questionnaires, screening tools, planning tools etc.

Risk Assessments

A risk assessment process should be applied to cGxP computerised systems in order to identify and mitigate potential risks to (1) patient safety, (2) product quality and (3) data integrity. Results identified through a risk assessment help to determine the validation strategy, the effort and time required, and allow better targeting of the validation activities to the highest risks.

The Risk Assessment should be revised during the Software Development Life cycle (SDLC) if the functionality, requirements or intended use of the system changes. The risk assessment activity should also be evaluated during system build-up as well as when implementing changes. Risk assessment tools for cGxP computerised systems are typically completed during the planning stage, specification stage and post qualification if a change or update is required.

Planning Stage

Initial Impact/Risk Assessment – during the planning phase to identify the level of impact and GxP relevance of the system/equipment. (Tools used: High-Level Risk Assessment).
Specification Stage

Functional or Quality Risk Assessment – during the specification phase – identify potential risks and possible mitigations to be to be introduced to the process. (Tools used: Quality Risk Matrix, (p)FMEA).

Changes to the System

Impact Assessment of Changes – as part of the change control process in the system operational phase. The following diagram defines the risk assessment steps within the system life cycle (Tools used: Impact Assessment Checklist, Change Control Procedures).

Quality Risk Matrix

A QRM is a risk assessment that identifies and manages the risk to patient safety, product quality and data integrity that relates to the systems processes. Risk scenarios or potential causes should be developed for each identified function or process step and then assessed for the impact on patient safety, product quality or data integrity. Risk mitigations and controls should then be introduced to address both medium and high levels of risk. The QRM requires three "assessments" in order to produce an estimation of overall risk (low, medium, high)

Assess Likelihood
Assess Detectability
Assess Severity

Traceability Matrix

A Traceability Matrix should be prepared as required in accordance with company and internal policy. It is also recommended by GAMP guidelines, ASTM E2500 and ISPE risk-based approach to validation. The matrix links the user requirements and specifications to the testing and validation activities. A traceability matrix illustrates that all user requirements are traceable to the verification/validation activity or vendor documents as relevant (FDS if applicable, design specifications etc.). Generally, individual organisations will have an approved template to work from. However, the URS structure can form the basis of the template, with additional columns added to document the test/verification method, reference documents (such as FDS and vendor specifications and design documents)

General Requirements

Configuration Identification

Software and hardware packages should be identified by a unique product identifier and a version number. For the software end-user, the parts of an automated system that are subject to configuration management should be clearly identified. The system should therefore be broken down into configuration items. These should be identified at an early phase of development so that a complete list of configuration items is defined and maintained. The application-specific items should have a unique name or version ID. The depth of detail when specifying the elements is decided by the needs of the system, and the organisation developing that system.

Requirements for the User ID and Password

User ID: The user ID of a system should have a minimum length agreed with the customer and should be unique within the system.

Password: A password should always consist of a combination of numeric and alphanumeric characters. When setting up passwords, the number of characters and a period after which a password expires should be stipulated. The structure of the password is normally selected to suit the specific customer. The configuration is described in the section Security Settings of Password Policy.
Criteria for the structure of a password are as follows:
Minimum length of the password
Use of numeric and alphanumeric characters
Case sensitivity

Audit Trail

The audit trail is a control mechanism of a system that allows all data entered or modified to be traced back to the original data. A reliable and secure audit trail is particularly important in conjunction with the creation, change or deletion of GMP relevant electronic records. In this case, the audit trail must archive and document all the changes or actions made along with the date and time. Typical contents of an audit trail must be recorded and describe the procedures "who changed what and when" (old value/new value).

Uninterrupted Power Supply

An uninterruptible power supply (UPS) is a system for buffering the main power supply. If the power supply fails, the battery of the UPS supplies the required power. When the power supply returns, the UPS battery stops supplying power and is recharged. Some UPS systems provide the option of main power supply monitoring in addition to the buffering function. They guarantee an output voltage at all times without interference voltages. UPS systems are necessary so that process and audit trail data can continue to be recorded during power failures. The design of the UPS must be agreed with the system user and must be specified in the URS, FS or DS. The following points must be considered:

> Energy requirements of the systems to be supplied
> Power of the UPS
> Required duration of UPS buffering

The energy requirements of the systems to be buffered decide the size of the UPS. A further selection criterion is the priority of the systems. Systems with high-priority include:

> Automation system (AS)
> Archive server
> Operator station (OS) server
> Operator station (OS) clients
> Network components

Field devices that generally have relatively high energy requirements may also be included in the buffering depending on the power of the UPS. This must be decided in consultation with the system user and related to the classification of the process. Whatever is decided, it is important that the systems for logging data are included in the buffering. The time at which the power failure occurred should also be recorded. The use of UPS systems involves the installation of software. This should be installed and configured on the PC-based computers of the process control system to be buffered. The setup should also account for:

- Configuration of the power failure alarms
- Stipulation of the time before the PC is shut down
- Stipulation of the time during which UPS buffering is provided

The automation systems (AS) must be programmed so that the process control system changes to a safe state after a selectable buffer time if a power failure occurs.

Types of UPS

Due to the different requirements of the various devices involved, three classes have established themselves as stipulated by the International Engineering Consortium (IEC) in product standard IEC 62040-3 and the European Union EN 50091-3:

Offline UPS

The simplest and least expensive UPS systems (according to IEC 62040-3.2.20, UPS class 3) are standby or offline UPS systems. They protect only against power outages and brief voltage fluctuations and peaks. Undervoltage and overvoltage are not compensated for. Offline UPS systems switch to battery supply automatically if there is overvoltage or undervoltage.

Line-interactive UPS

The way in which line-interactive UPS systems (according to IEC 62040-3.2.18, class 2) function is similar to standby UPS systems. They protect against power outage and brief voltage peaks and can compensate voltage fluctuations continuously using filters.

Online UPS

Double conversion or online UPS systems (according to IEC 62040-3.2.16, Class 1) count as genuine power generators that continuously generate their own line voltage. Connected consumers are therefore supplied permanently with line power without restrictions. At the same time, the battery is charged

Software Source Code Review

For GAMP Software Categories 4 and 5, source code review is advised unless the supplier has evidence of the same available for review. As part of Good Automated Manufacturing Practices, reviews should be completed as part of the development life cycle. If a source code review is not completed a justifiable rationale should be documented in an applicable document such as a validation master plan.

Calibration

A key part of any qualification is to confirm that the equipment is fit for the intended purpose. Each piece of equipment will have a defined operating range. For example an oven may have an operating range of 20°C to 100°C ±5°C. However, the process window may only require a temperature range of 30°C to 60°C. In this instance a calibrated range of 20°C to 70°C would suffice. However, if the process window or the temperatures at which product was manufactured ranged from 20°C to 100°C this would present a problem as it falls outside of the equipment qualification range when the calibration tolerance is taken into account.

Deviations

A deviation can be simply described as an unintended event which causes a test or verification to fail to meet expected acceptance criteria. Each company or organisation should have a procedure detailing the management of deviations. It is critical that all deviations are identified, investigated and evaluated for their impact on product quality, the risk/impact to the patient and the impact on the qualification or validation. The basic components to a deviation are listed below:
- Deviation Description - provides the page and section of the deviation and an overall description e.g. document generation error, operator error, machine crash etc.
- Potential impact on product – does the deviation impact the product?
- Potential impact on validation/qualification – will the validation have to be repeated in part or in full?
- Investigation – DMAIC, RCA, fishbone diagram, 5Ws
- Root Cause - what is the concluding root cause?
- Planned Resolution - what actions are required to be implemented?

Deviation Resolution (Actions completed) – were all the actions in the planned resolution implemented? What is the final result? Have the actions been effective?

Requalification

Over the lifetime of a piece of equipment, the need to requalify may arise. Therefore, any proposed change to equipment or a process must be assessed to see if the validated state will be impacted. It is therefore critical to understand clearly the nature of the change(s). Some scenarios where requalification of equipment may be required include:

- Major equipment repairs
- Moving equipment
- Changes to the upper and lower operating limits of the equipment
- Upgrading of software
- Hardware upgrades or changes
- Changes in performance and/or defect levels

After assessing any proposed changes based on the reasons listed above, a determination of the level of requalification is required. This may be limited to a partial requalification (addendum) or it may require a full requalification.

CHAPTER 4
PROCESS VALIDATION

Introduction

This chapter provides an introduction to process validation for medical devices. Process validation is a statutory and regulatory requirement for the manufacture of medical devices. Per FDA 21 Code of Federal Regulations process validation is a regulatory requirement of Good Manufacturing Practices (GMP) for both pharmaceuticals (21 CFR 211) and medical devices (21 CFR 820). In addition to the regulatory drivers, process validation is a requirement in order to obtain certification to international standards issued by many notified bodies. (E.g. ISO 13485 Medical Devices – Quality Management Systems, ASTM E2500-Standard Guide for Specification, Design, and Verification of Pharmaceutical and Biopharmaceutical Manufacturing Systems and Equipment etc.)

Traditional and New Approaches to Validation

Historically, process validation involved the testing and verification of all aspects of a process. While this may seem appropriate, it must be understood that in order to test/verify all aspects of a process, and for it to hold weight, this activity must be documented and recorded. In this respect, an "all aspects" approach to process validation can be burdensome to resources. The traditional approach largely used the V-Model which set out a sequence of deliverables that should be completed. The use of risk assessments were limited as all requirements of a system were tested and qualified.

In recent years, a risk-based approach has been increasingly endorsed by regulatory authorities and hence adopted by medical device manufacturers. One such standard is the ASTM E2500. As the title suggests, it is primarily used within pharmaceutical and biopharmaceutical industries; its principles and core approach can be adopted by medical device manufacturers also. ASTM E2500 was designed to make the implementation process for GMP systems and validation more cost-effective. It aims to achieve this based on scientific and risk-based principles, focusing on the risk to the patient. However, at just a five-page document, ASTM E2500 lacks the detail required in order to meet regulatory expectations. While different terminology and philosophies exist they do not change the regulatory expectations relating to validation.

Both approaches exhibit common elements which include:

- o Good engineering practices
- o Planning
- o Requirements definition (URS etc.)
- o Design review
- o Change management
- o Documented testing and inspection

While many manufacturers may predominantly choose a particular approach, it is common to see elements of both approaches (traditional and risk based). Each individual company will shape its internal validation procedures to best suit the business needs of the company.

What Is Process-Operational Qualification (OQ-P)?

The ability of a process to produce product in accordance with pre-determined specifications under worst case conditions. PQ is only required if no worst case conditions are evident.

What Is Process-Performance Qualification (PQ)?

The ability of a process to consistently produce product in accordance with predetermined specifications under anticipated conditions (normal/routine conditions). Before considering process validation in further detail, it is important to look at the prerequisites and other supporting activities required. These are examined in the sections below.

Test Methods and Process Validation

It is important to consider test methods early on in the validation life cycle. Before you can begin to consider process validation, test methods should be understood and in place.

A test method is a process or an action used to verify that a product feature meets a predefined specification. Tests methods can be physical or analytical in nature. Test method validation should be completed in advance of process validation to allow the proper assessment of process and product outputs meaning it is often a pre-requisite to process validation.

Examples of test methods include simple visual inspection by microscope, measurement of a dimension with a callipers or measurement of a dimension using an automated optical inspection system. Some test methods will involve MSA (Measurement System Analysis) studies, for example, a measurement of a dimension by an operator using a microscope. In contrast, a test method to determine organic residuals would require an analytical test method validation.

The equipment must be qualified (installation qualification and operational qualification) before the method is validated. Remember – testing completed in contract laboratories or specialist services also require validation! Test methods are critical to the success and integrity of process validation as they assess the outputs. E.g. what are the dimensions, physical attributes or chemical properties of the product and how do they conform to specifications?

Stages of Process Validation

The three stages of process validation include:

- Process Design
- Process Validation
- Continued Process Monitoring

The commercial manufacturing process must be established during the process design phase. Some typical activities include:

- Definition of process inputs
- Effects of inputs
- Process outputs – CQAs (critical quality attributes)
- Establishing process windows
- DFMEA /PFMEA (design/process)

Design Control procedures should be developed to allow proper management of the process design stage. At the process design stage, the business must define the manufacturing process. This often involves liaising with vendors and Subject Matter Experts (SMEs). The process qualification stage looks at the validation of process design to confirm process is operating as intended and is capable of consistently producing product to meet quality requirements. Finally, stage 3, Continued Process Verification provides ongoing assurance through regular testing and verification to ensure the process is in control. Stage 3 is often referred to as In-Process Control or In-Process Testing (IPC/IPT). This data provides feedback to engineers allowing them to trend the performance of output data. This can identify deficient equipment, changes in wear tooling etc.

Fundamentals of Process Validation

The most important point when it comes to validation is that validation is neither exploratory nor investigative. Equally, it is not an engineering study. If you are ready to validate a system or process, all of the groundwork must be completed. This means critical parameters must be defined and documented, with technical rationale on why such parameters are critical etc. This body of work is typically done during a process development study or protocol. Process validation is confirming that a process is capable of consistently manufacturing product under anticipated conditions. Remember, validation should be representative of the commercial process, so any issues in process validation will be repeated in commercial manufacturing.

Consistency, a core principle of process validation, is typically demonstrated by producing three batches/runs for a Process Performance Qualification (PPQ). These batches should be representative of normal production i.e. the size of the batch should be typical of commercial volumes. The PQ study should be executed at nominal conditions, (often termed "anticipated conditions") essentially referring to a controlled environment. Controlled material and controlled parameters (CPPs) are required. Nominal settings should be selected for PQ.

Process Validation and Dominance Factors

The concept of dominance is a term used to describe the "influential" or "dominating" effect on a system or process. Typical examples include the injection moulding process, and packaging process. For example, an injection moulding process can be said to have material as a dominant factor. Batch-to-batch differences of resin or raw material may cause a change to outputs such as the dimensions of a product or component. If dominant factors cannot be identified or understood a "Designed Experiment" (DoE) technique can be used to properly determine them.

Dominance can be categorised into five sections: (1) setup dominance (2) time dominance (3) worker dominance (4) information dominance and (5) component dominance.

Setup Dominance

Setup Dominance - The Process or equipment relies principally on a procedure or process setup. Process should be a stable one "set-up".

Examples include ovens and package sealers. With regard to the oven, the setup would generally be controlled by a recipe or program. This program would be selected by the operator through the Human Machine Interface (HMI). The setup with the correct version of the recipe that contains the desired temperatures, times and pressures is therefore a critical input to the process. With regard to the packaging machine (blister packaging), the correct setup for the tooling and program are critical inputs. If setup dominance is significant, it is best practice to have three separate set-ups/changeovers in the Performance Qualification (PQ).

Time Dominance

The Process or equipment is subject to changes over time (drift over time in temperature, solvent cleanliness, tool wear etc.) The process may need a schedule of process checks and adjustments to ensure process consistency.

Examples include CNC Machinery (tool wear) or aqueous based cleaning systems. The tool may only be able to manufacture 1000 parts before defects or quality issues are encountered. If time dominance is significant, three time points or cycles of expected variation should be made e.g. three points in the cycle (start, middle and end) or three points in a shift (start of shift, middle of shift and end of shift).

Worker Dominance

For worker dominance, the process requires operator experience and skill. Examples include manual or hand finishing. If dominance is significant, ensure there are a minimum of three operators involved in the manufacturing/ activity.

Information Dominance

With information dominance, the process or equipment requires the transmission and/or analysis of information. Examples include LIMS, MRP and ERP systems. A minimum of three information transmissions in the PQ should be completed.

Component Dominance

The process is influenced by the variability of the input materials and/or components. It requires robust inspection and sorting procedures as well as process adjustments. When component dominance is significant, ensure there are a minimum of three component/raw material batches in the PQ sampling plan. If component dominance is significant, this can be mitigated by including the material/component variation in "worst case" testing as part of the Operational Qualification Process (OQ-P)

Process Operational Qualification (OQ-P)

During the Operational Qualification-Process (OQ-P) study, worst-case process conditions are normally employed. This may be worst case temperatures, speeds, feeds etc. The OQ-P should challenge the manufacture/processing of product at the limits of the processing window. If no worst-case conditions exist, then an OQ may not be required and only a performance qualification is required.

A family or matrix approach is often used where similar products are to be validated. A particular product size of product configuration may be selected to represent the worst-case product. Therefore, by qualifying the worst case, all other products within that family of products would be considered validated. However this approach must be clearly documented and technical rationale provided in advance of any qualification activities. This can be addressed in a validation plan or within a protocol.

Protocol Approval Check list

The validation protocol is the means by which objective evidence is documented and gathered. The validation protocol is therefore a critical document. It should clearly set out the approach to the validation, detailing methods, tests and verifications to be completed and the acceptance criteria that applies to such tests and verifications. Remember, a validation document is a legal and regulatory document and can be subject to detailed scrutiny. Below are some suggested general checks to apply when writing validation protocols.

Author:
- SOP available - Protocol conforms to validation procedure.
- Ensure item numbers and batch size are correct.
-Test methods are correct.

SME Reviewer:
- Is the protocol number correct?
-Review content of protocol for accuracy and completeness.
- Protocol conforms to validation procedures.
- Procedure and evaluation table are appropriate and correct.

Engineering:
- Review content of protocol for accuracy and completeness.
- Specifications and operating parameters are correct.

QC / Laboratory:
- Review content of protocol.
- Raw material specifications are in place.
- Finished product specifications are in place.
- Testing and sample size is correct.

Quality:
- Review content of protocol.
- Protocol conforms to SOPs.

- Evaluation and acceptance criteria are appropriate.

Process Performance Qualification

The purpose of the PPQ is to demonstrate the capability of the process to consistently manufacture product to pre-determined specifications under normal operating conditions and defined parameters.

Key principles of Process Performance Qualification

Validation is confirmation, so process validation is confirming that a process is capable of consistently manufacturing product under anticipated conditions.

- Lots should be produced consecutively (in sequence)
- Lots must meet the acceptance criteria set out in the protocol
- The lot size should be reflective of the intended lot size and also take into account normal variation
- If a family approach or matrix approach is used, the product selection must be clearly justified and documented
- Execute under anticipated conditions; essentially this refers to a controlled environment. Controlled material, controlled parameters (CPPs)
- Nominal settings should be selected for PPQ

Yield Data (aka Process Yield Data)

Process yield is a term used in manufacturing to represent the overall process performance. Yield is most often expressed as a percentage of goods/passing products. It reports the percentage of compliant units, that is units or products that meet the product acceptance criteria (e.g. CQAs). The remaining "bad" units are classified as defects or scrap. In some manufacturing processes, rework is possible or permitted.

Yield data often forms part of the acceptance criteria for a validation. The overall process yield for each batch should be calculated and compared to the starting process weights or units to determine loss due to processing as it is common to lose material during processing.

Continued Process Verification

Once the initial validation is completed it is important that the system or process remains within the validated state, meaning that the system remains in a state of controlling process systems that capture information and data about the performance of the process. The use of statistical trending techniques should be considered. Data analysis of process and product should also include trending of raw materials, components and finished product. The purpose of process monitoring is to ensure critical parameters remain within control limits. It also helps to identify increasing variability or instability within the process which can then be investigated. All processes must have an upper and lower limit. If a process parameter only has a one-sided limit, then provide rationale in the OQ protocol to justify why a one-sided parameter window is acceptable. This requirement is not applicable to parameters that are set points.

Revalidation (or Maintaining a Validated State)

Revalidation is sometimes required if the original validation is no longer valid or representative of the process. Some instances where revalidation must be considered include changes to the process that can affect the product quality or efficacy, a removal, or the addition of a processing step or transfer of the equipment to a different location. In many companies an impact assessment is conducted if there is a proposal to modify a manufacturing process. Some changes may not require any validation while others may require a verification run.

When changes are proposed to the validated state of a process, the proposed changes must be fully understood in terms of the impact to product quality and the validated state. A risk assessment should be conducted to determine risks and appropriate mitigations.

Other Scenarios – Maintaining a Validation State

Line Addition-Product (New Product or Product Transfer)
This may be required if a new product has been introduced but uses the same process(es) for manufacturing. Typically this can apply if a new size has been introduced. For example, a new size of surgical blade.

Line Extension

A line extension commonly refers to a scenario where the product/process is different or considered outside the existing range or processing parameters.

Note: The impact on the validated state for line additions and line extensions should be assessed formally and documented. A line extension may require a new validation or addendum to the existing validation, whereas a line addition to add a new product may be within the scope of existing validations.

Acceptance Criteria

The acceptance criteria contained in validation protocols are normally based on established product specifications. For example, a contact lens manufacturing company may produce a lens with a diameter of 18mm ±0.2mm. The product produced during a process validation must be inspected to record the diameters of lenses being manufactured. Disposition of product is based on the product specification and determines if the product feature measured receives a pass or fail.

In addition to product specifications, it is common to have acceptance criteria such as yield, and OEE. The acceptance criteria for these conditions are normally driven by an internal company procedure or alternatively can be detailed in the validation plan or protocol study.

Validation Strategies

A Family Approach (a.k.a. Bracketing, Matrix Approach) to validation is often used where a variety of similar products are manufactured using the same equipment. For process validation, a product that is representative of the family or group of products may be selected. Alternatively, a 'worst case' product may be selected as it presents the greatest challenge to manufacture to product specifications.

Principles of Worst Case Selection

Worst Case is a particular condition, set of conditions, and/or set of process parameters, generally made up of processing limits. Worst case conditions present the greatest chance of process issues or the greatest chance of failures due to product quality. Worst case conditions are used at the OQ-P stage to provide the greatest level of challenge, however, this is outside of normal operating conditions.

Requalification

During the lifetime of a process or piece of equipment, the need to re-qualify may arise. Such need should be assessed according to a validation procedure. Generally, the same tools used in the original validation can be re-applied to identify the need to re-qualify and indicate what requirements must be included.

The first step must be a review of the existing qualification, as changes may not impact the validated state, or may only require a limited requalification. For example, moving a piece of equipment may only require requalification of the utilities such as compressed air or process water if the operation of the equipment is not impacted by the movement and re-siting.
Some examples where re-qualification may be required include:

Transferring a process from one plant to another plant
Changes to the process settings which may impact the product quality
Changes to the design of the product
Changes to manufacturing aids (e.g. cleaning agents, jigs and fixtures)

Process Validation - Examples of Deficiencies

Identification of Critical Process Parameters

No procedure in place to define or identify Critical Process Parameters (CPP). Not documenting from where these CPPs would be obtained when writing process validation protocols.
In a drying process for an oven, no rationale was documented as to why time and pressure were not considered CPPs. Temperature was only considered a CPP.

Reproducibility

For a concurrent validation, three separate reports were drawn up and each batch had been concurrently released. However, no summary report was generated to assess the reproducibility of the process.

Risk Assessment

No linkage between the validation protocol and the various controls in the manufacturing process that had been identified in the risk assessment as important for risk mitigation.

Case Studies on Process Validation

The following case studies review key considerations when it comes to process validation. For your benefit, we have focused on the critical elements which include:

1) General principles of process validation
2) Dominance factors
3) Parameters and settings
4) Other considerations

Case Study - CNC Grinder–Performance Qualification (PQ)
System Overview

A Computer Numerically Controlled (CNC) Grinder uses grinding wheels to machine complex geometries or modify surfaces. As with any process validation, it should factor in when to influence the process (dominant factors). The machine settings are also another factor which need to be considered for process validation. Machine settings should also have a tolerance or +/- value associated with them. However, with CNC machines, as they are numerically controlled, parameters such as spindle speed tend to be accurate. In the other case studies, we will see other parameters such as temperature and how tolerances are associated with settings.

General Principles of Process Validation applied to a CNC Grinder

As the spindle speed of the machine is "set-point" only and is known to be accurate, an Operational Qualification (OQ) is not required as there are no 'worst case' settings. All system settings are set-points and the only variable is 'wheel life'. As the wheel is used to manufacture/grind parts, over time the wheel's grinding performance decreases, therefore, the wheel has a life expectancy. Wheel life assessment should form part of the performance qualification. A performance qualification manufactures components at start, then middle and at the end of wheel life. First article inspection of (FAI) should be performed at the start of each run in order to confirm the "first off" is according to specification and that the machine setup is correct.

Dominance Factors

For a CNC grinder, appropriate dominant factors include time and setup. Time is a factor as the wheel life changes over time. Setup is also a factor as the tooling and fixturing must be setup accordingly and this can be a source of variation that must be challenged as part of the PQ.

Parameters and Settings

Coolant temperature, grinding feed rate, cutting dwell time and spindle speed all have the ability to affect performance and product quality, therefore they can be considered controlled parameters. Coolant temperature should have a tolerance specified in the validation protocol e.g. +/- 2 degrees.

Other Considerations

The incoming raw materials e.g. mild steel bar stock, should confirm to an acceptance criterion of Certificate of Analysis.

Case Study – Cleanline Operational Qualification and Performance Qualification (OQ-P/PQ)

System Overview

Cleanlines are used to clean fixtures, equipment parts and products within the medical device industry. Cleaning processes can be loosely divided into intermediate cleaning processes and final cleaning processes. A higher acceptance criterion is applied to final cleaning processes. Cleanlines are aqueous or solvent-based systems. Aqueous systems typically use deionised water and detergent, followed by DI rinsing and dryings steps. Solvent systems are also effective at removing grease and oils from parts and components.

General Principles of Process Validation Applied to a Solvent Based Cleanline

A cleaning process is made up of: 1) cleaning parameters such as temperature, time and ultrasonics 2) manufacturing agent, in this case solvent and 3) worst case conditions. The heating or cooling of liquids (although temperature controlled by a PLC-temperature probe) is subject to drift. This drift can vary on a given day, or due to the use of the equipment. Therefore, it is important to quality-check an operating range for the equipment and the cleaning process. This operating range is qualified by the execution of an OQ-P, Operational Qualification-Process. The OQ-P represents the limits at which products will be cleaned. The OQ-P aims to demonstrate that if the process settings are subject to drift in time or temperature, the cleaning process is still effective as it operates within the validated operating range. This operating range is also known as the process window.

The following strategy should be applied for the Operational Qualification- Process and Performance Qualification of a cleaning system:

OQ Low, 1 lot
OQ High, 1 lot
PQ nominal, 3 lots
PQ nominal rework, 1 lot

Full loading conditions should be applied for all cleaning run/conditions. This means that the basket or "carrier" that holds the pieces while in the washer, is full to the normal, anticipated capacity. This is to create the conditions as they are intended during commercial manufacturing.

Dominance Factors

The dominant factors for a cleaning process are considered to be time based. As the system is used, the water/solvent in the tank gets more "soiled" as time moves on. It essentially gets dirtier; therefore, there may be a difference in the cleaning performance of the system at the start of the day (clean water with fresh detergent) as opposed to the end of the day, after a full shift of cycles (dirtier conditions).

Parameters and Settings

Parameters include rinse temperatures, times and ultrasonic frequencies. Some equipment may be fitted with in-line conductivity meters and/or pH probes which indicate the "cleanliness" of the solution. If a drying stage is incorporated into the system, this will also have temperature and time settings and tolerances. Note, that some basic systems may only have fixed ultrasonics which cannot be varied.

Other Considerations

An important consideration when conducting cleaning process validation for medical devices is the manufacturing agents (greases, oils etc.) that are used up-stream in intermediate steps or processes. The chosen cleaning detergent or cleaning solvent should be selected in a systematic way to ensure it is effective in removing the "soiled" material. Some key points to assist in this process are:

Map out the complete manufacturing process identifying the stage at which cleaning is completed.

☐

CHAPTER 5
PACKAGING VALIDATION

The content of this chapter provides an outline of six key stages of the packaging validation life cycle. In order to describe the packaging life cycle in a structured and manageable format, the process can be sub-divided into 6 stages. It must be noted that an individual company will have its own interpretation of the required stages with different terminology or specific requirements. However, the intent of any approach should broadly align.

Stage 1 - Design and Development of Packaging
Stage 2 – Material(s), Equipment and Process Technology
Stage 3 - Material Performance and Suitability Testing
Stage 4 - Stability Testing
Stage 5 - Packaging Performance Testing
Stage 6 – Packaging Validation

Blister packing is a process where pre-formed plastic packaging (blisters) manufactured using a cavity of a defined shape and size are used to pack a range of items including various personal tech-ware goods, foods, pharmaceuticals and medical devices. The primary component of a blister pack is a cavity or pocket which is made by forcing a material, usually a plastic into the cavity under vacuum and at deformation temperature. The resulting blister is often closed with cardboard, paperboard, foil or plastic depending on the product. The combination of the blister and lid or lidstock helps to protect products from the environment such as microbial contamination, humidity and foreign matter.

```
                    Lid
    ▄▄▄▄▄▄▄▄▄▄▄▄▄▄▄▄▄▄▄▄▄▄▄▄▄▄▄▄▄
    █ ▐                     ▌ █
    █ ▐      Blister        ▌ █
    █ ▐      (Cavity)       ▌ █
    █ ▐                     ▌ █
    █ ▐▄▄▄▄▄▄▄▄▄▄▄▄▄▄▄▄▄▄▄▄▄▌ █
```

Figure: Simple representation of a blister cavity and lid or lidstock.

Most blisters made from plastics will provide good protection to the inner product. Often, it is the lid or lidstock that is the weaker of the two. Some lid materials are prone to tearing or puncturing and degradation over time. In addition, the area where the lid is bonded to the blister is a point of interest and must be inspected adequately to ensure a proper seal integrity is achieved. Not only is the movement of the product and force of the product a risk factor in damaging the package, but equally the design of the system should also account for handling and a degree of inappropriate handing as a safety factor.

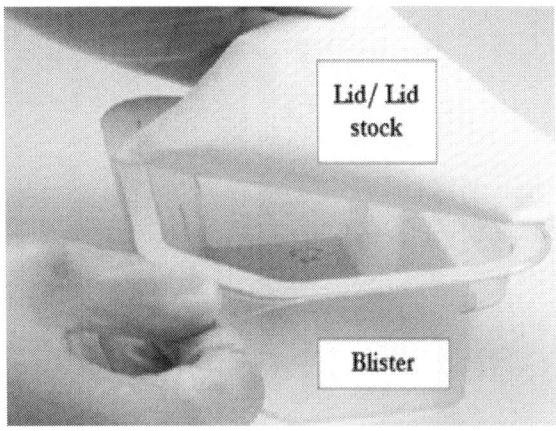

Some manufacturers offer breathable-type seals which reduce the risk of condensation forming on the inside of the pack due to temperature and humidity differences that can occur during shipping or storage.

Stage 1 - Design and Development of Packaging

The design of medical packaging is equally as important as the products they contain. They ensure products are kept clean, sterile (if applicable) and essentially make them safe and effective when used. They need to serve the requirements of the regulatory bodies and international standards but also provide ease-of-use to patients and users. Packaging systems can include lids (lidstock), pouches, bags, trays and blisters that are used to contain the drugs and medical devices.

General Requirements for Design and Development

Validated Test Methods: Test methods are used to verify the outputs of manufacturing processes. In the case of packaging, some examples of test methods include seal strength.

Robustness of Process: Tests selected for use must adequately address the robustness of the packaging system and process being tested. The rationale for the tests selected for use must be documented in the development protocols.

Design: The design of the blister and lid are properly scoped out, documented and controlled.

Sample Size: Test sample sizes must be based on statistical rationale. The rationale should be documented in the development protocols.

Key Requirements of Medical Packaging

The specific requirements on a given packaging system depend on the classification of medical device. Medical devices are classified based on the application of the devices and the level of risk associated with their use. In general, the level of risk is understood to increase with the (1) duration of use (2) level of invasiveness. Therefore, the packaging requirements for a particular product must be specified correctly and in keeping with the intended use.

High-density polyethylene or HDPE is a popular material that aids in the prevention of microbial contamination. However, the contents and environment must meet microbial standards if a sterile product is being manufactured. The effectiveness of HDPE is based on the amount of very fine filaments and their random orientation makes it an effective barrier.

During the product design phase, the packing configuration must be selected. This selection impacts the equipment and manufacturing technology required in order to deliver the designed barrier system. Most products in a modern manufacturing facility will require a medium to high-volume manufacturing. This is typically driven by market need. If high volumes are not anticipated, a high cycle time may be required to allow the bare minimum number of machines to deliver the throughput required. Storage and shelf life of medical products is also a key concern for the patient and user. In particular, HDPE lidstock can maintain sterility for up to five years. The process of sterilisation involves the controlled release of gases or steam that "penetrates" the packed product but can then quickly escape from the packaging and leave the product sterile and unaffected. Materials must be suitable for the type of sterilisation process.

Inputs

Packaging materials must be compatible with the chosen method of sterilisation, therefore, during the design and development stage, the type of sterilisation must be selected carefully. This is based on regulatory requirements and the capability of the product as well as primary and secondary packaging materials. The requirements with regard to acceptable foreign matter, visual defects, seal strength and integrity criteria must also be developed to form what will be acceptance specifications.

User Requirement Specification (URS) A URS is a requirements document that specifies the intended use of the equipment along with specific operating and process requirements unique to a particular product. The scope of a URS document can be sub-divided into three main sections: (1) installation requirements (2) operational requirements and (3) process requirements.

Installation Requirements: These relate to the type of facility and space available. The footprint and weight of the packaging machine may be a factor for some premises. The customer may also which to procure a packaging solution that is mobile.

Utilities: Typically, electrical and pneumatic supply options are specified in a URS and the machine manufacturer must confirm that the equipment can be successfully operated with the available utilities onsite.

Operational Requirements: Operational requirements may be specific to a particular product or family of products. Some typical examples include:

- Equipment must reach operating parameters from standby within 15 minutes of cold start-up
- It shall be possible to operate the equipment with one person
- Equipment shall be capable of processing a minimum of 20 blisters per minute

Process Requirements: Process requirements can also relate to a specific product or packaging configuration. For example, "the seal shall meet the minimum width specification of 6mm". While other process requirements are more generic:

- No smudge marks, burn marks or tool marks shall be visible
- Seal areas must be free from creases or any other defects

Outputs

Outputs are essentially the packaging features or attributes that need to be validated and found to be within acceptance levels. For blister packaging, the typical requirements are listed below:

Integrity of sterile barrier
Tensile strength of seal and delamination
Seal width
Cosmetic requirements

Outputs must be considered from the very beginning of a packaging project prior to packaging validation. The choice of materials, equipment and technology can all impact the outcome of the final packaged product. Choosing the wrong materials may cause long delays when issues are discovered or encountered. Choosing the wrong technology, equipment or process may not provide the required capability or necessary quality, especially for regulated products (e.g. medical devices and sterile products).

Sterile Packaging

Packaging materials must be compatible with the chosen method of sterilisation. Sterilisation methods include ethylene oxide, electron-beam, gamma, electron-beam, steam (under controlled conditions) to name but a few. Packaging must provide a high microbial protection and breathability along with acceptable levels of tear resistance and durability.

For sterile medical devices, regulation requirements per EN ISO 13485:2013 include:

"Devices delivered in a sterile state must be designed, manufactured and packed in a non-reusable pack and/or according to appropriate procedures to ensure that they are sterile when placed on the market and remain sterile, under the storage and transport conditions laid down, until the protective packaging is damaged or opened."

"Devices delivered in a sterile state must have been manufactured and sterilised by an appropriate, validated method. Devices intended to be sterilised must be manufactured in appropriately controlled (e. g. environmental) conditions."

"Packaging systems for non-sterile devices must keep the product without deterioration at the level of cleanliness stipulated and, if the devices are to be sterilised prior to use, minimize the risk of microbial contamination; the packaging system must be suitable taking account of the method of sterilisation indicated by the manufacturer."

"The packaging and/or label of the device must distinguish between identical or similar products sold in both sterile and non-sterile condition".

Stage 2 – Material(s), Equipment and Process technology Supplier Requirements

Above all, materials must be suitable for the intended use and classification of medical device. Most vendors will operate to a quality management standard such as ISO 9000 or ISO 13485. When dealing with vendors and external suppliers of packaging materials, all medical device manufactures should adopt a vendor approval procedure to specify the supplier requirements.

Materials

Polyvinyl Chloride (PVC)

PVC is a low cost material and suits blister packaging due to its suitability to thermoforming. However, it offers limited barrier protection against moisture and oxygen ingress. PVC blisters provide good protection for physical pharmaceutical solid dose tablets and caplets. PVC sheet thickness is typically between 200μm to 300μm depending on the cavity size and shape. PVC does not provide the highest protection with regards to water vapour ingress. This can be improved by laminating processes using PVDC. To meet suitability for use requirements, PVC formulations need to meet standards such as the US Pharmacopoeia 661, FDA 21 CFR and local regulatory requirements.

PVDC

Polyvinylidene chloride or PVDC is often combined with PVC film by using a lamination technique in order to gain better moisture and oxygen barrier performance. PVDC coated blister films are the most common and prevailing barrier films used for pharmaceutical blister packs.

Cyclic Olefin Copolymers (COC)

Cyclic olefin copolymers (COC) or polymers (COP) can provide moisture barriers to blister packs, typically in multi-layered combinations with polypropylene and polyethylene. Cyclic olefin copolymers have good thermoforming properties even in deep cavities, leading some to use COC in blister packaging as a thermoforming enhancer, particularly in combination with polypropylene or polyethylene.

Lidstock

As previously mentioned, high-density polyethylene or HDPE is a preferred barrier material as it provides excellent protection against water and oxygen ingress. It also can be manufactured to provide good tensile strength offering protection to the product.

An alternative to HDPE lidstock is foil based lidstock. Foil based lidstock can be designed to be heat sealed to polymer blisters such as polypropylene and it suitable for steam sterilisation at high temperatures (over 100°C. Foil based lidstocks also achieve good tensile strength protection for the packaged product.

Equipment and Process Technology

This section describes two approaches to blister packaging of medical devices. The first approach adopted by some manufacturers (especially at initial launch where volumes are relatively low) is for the project or packaging engineer to select a simple manual process. Such a process may consist of the manufacturer receiving pre-formed blisters from a vendor or supplier. The manufacturing only then needs to worry about placing the contents in the blister and sealing it with a lid. The preformed blisters are loaded into each position, the product is placed in the cavity, a lid is placed on the top side of the blister and the sealing process can begin. A more advanced method of blister packaging involves more automated equipment that completes both the forming of the blister and sealing of the lid-to-blister. Blister sealing is typically completed through pressure heat transfer over a short period of time. Controlled parameters include:

       Seal Temperature
       Seal Time
       Seal Pressure

The seal settings for the above parameters must be determined during process development prior to the commencement of any process validation (OQ and PQ).

Stage 3 - Material Performance and Suitability Testing

The material performance and suitability is demonstrated through testing. This often involves the development of technical reports or the execution of engineering studies to gather the evidence and appropriate rationale to support suitability for use. Prior to any functional or physical testing, test methods need to be validated in advance to ensure they are fit for purpose.

Test Methods

Test methods are used to measure both variable and attribute data that is generated as part of a validation. Variable outputs refer to data that is parametric in nature or continuous. Non-variable data, also known as attribute data is non parametric (such as pass/fail visual inspection)

Variable Outputs

Test method validation must address the following parameters for test methods with variable output:

Accuracy
Precision
Range
Resolution

Attribute Outputs

Test method validation addresses the following parameters for test methods with attribute output:

Effectiveness
Probability of false alarms
Probability of misses

Prior to any test method validation, the test method itself (SOP/ procedure) should be available in draft form. The test equipment along with any software should also be qualified and fit for purpose.

Seal Width Measurement

Seal width measurement is a process output that helps to determine the seal integrity post blister sealing. The variable data can be used to monitor the seal quality of blisters and identify any changes in the process that might affect the barrier system.

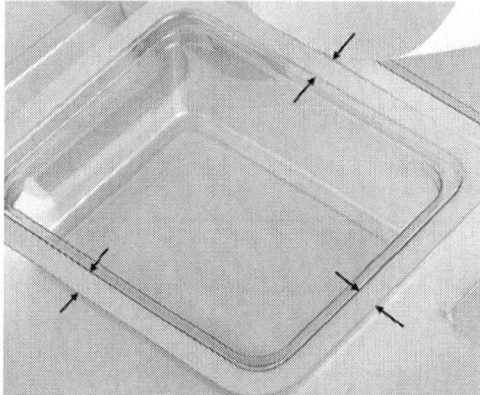

Figure: Seal width

Tensile Testing

Seal strength and seal integrity are critical outputs of the blister sealing process. The worst case sterilisation condition for a material is dependent on the material and the method of sterilisation. However, sealing and subsequent sterilisation at upper and lower worst case conditions should be completed. Tensile testing should be completed in accordance with ASTM F88 or another recognised standard.

Dye Penetration Testing

This test is designed to evaluate the integrity of a sterile barrier system. A blue dye is syringed into a sealed blister with the lidstock intact. The dye should then be allowed come into contact with seal for a defined period of time. Dye penetration testing should be completed in accordance with ASTM F 1929.

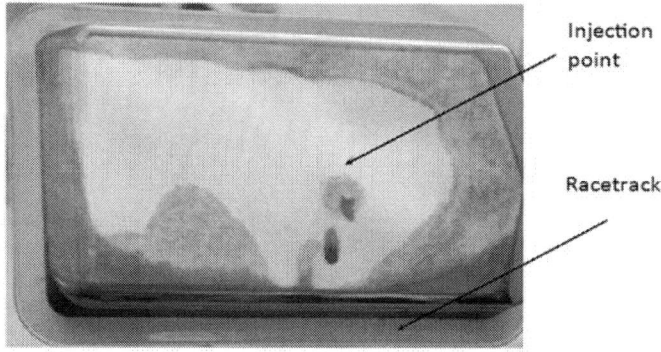

Figure: Dye penetration test showing the blister "seal width" racetrack clear of dye indicating an adequate seal.

Packaging Inspection (Cosmetic)

Packaging inspection should be completed pre and post sterilisation in order to ensure product being sterilised is not damaged or compromised prior to sterilisation. Some typical cosmetic checks are listed below. Packaging inspection requirements should be detailed in an approved specification.

- The entire package must be free of foreign material
- The maximum permitted number of inclusions should be specified along with the max size of the inclusion in mm^2
- No smudge marks or burns
- No tool marks
- No voids or bubbles
- No pinholes or tears

Note: For dye penetration, there should be no evidence of penetration across the complete width of the seal.

Bioburden Testing

Bioburden testing will be performed on devices pre-sterilisation to determine the levels of viable organisms that are naturally present on the product or introduced artificially.

Bacterial Endotoxin

Bacterial endotoxin testing is performed in order to test for the presence of bacterial endotoxins of the product. An endotoxin detection test involves testing the liquid sample (or the sample extract) with Limulus Amebocyte Lysate (LAL). LAL is an aqueous extract of the blood cells of horseshoe crabs. LAL forms a clot or changes in colour, depending on the technique used, in the presence of bacterial endotoxin. The test sample is compared to a standard series of Control Standard Endotoxin (CSE) dilutions.

Biocompatibility

Testing is necessary in order to evaluate that the material is biocompatible and appropriate to the intended use of the material/ finished product. Testing ensures the safe application of the device if it is in contact with the body or in used invasively.

Additional Testing

The below testing is normally not a requirement of the packaging validation itself, however, testing may need to be completed during the packaging system development stage or process development in order to ensure the design requirements are met:
Puncture resistance testing is used to measure the toughness of a material punctured via a standard method to determine the relative ability to experience a puncture failure.

Abrasion resistance is a test to determine failure due to rubbing of the product or part of the
package against the primary sterile barrier and the potential to create pinholes or other failures.

Stage 4 - Stability Testing

The purpose of Stability Testing for packaging is to verify that the packaging materials meet requirements over time. Examples of requirements include sterility, functionality, safety, efficacy and visual appearance. Stability testing is also known as "ageing." Stability testing is required for all packaging materials, including blisters, films (lids), cartons and so on. The project team in conjunction with the packaging engineer determines the appropriate testing and time-points and conditions required during testing to verify the packaging system remains fit for purpose over the shelf-life of the product.

In addition to stability testing over real time, stability testing is also conducted in an accelerated manner. This is referred to as "accelerated" testing or "accelerated ageing".

Testing intervals should be determined based on the shelf life specification of the product in question. The first time point is always t=0 when packaging has just been completed. Sterilised and non-sterilised product should be tested at t=0. Testing must also examine the stability of labels to ensure they remain intact, the material does not crease and the artwork remains legible.

Stage 5 - Packaging Performance Testing

Performance testing the packaging system challenges the acceptability of the entire package system. Performance testing evaluates the interaction between the packaging system and the product in response to the stresses (events) imposed by the production processes and limits, sterilisation processes, storage and transportation environment. The testing is intended to demonstrate that the SBS and protective packaging are adequate to protect the product while maintaining sterility to the point of use. The worst case product configuration is used for this testing.

The following should be considered for inclusion in engineering trials or technical reports:
- Product or representative product is necessary for performance testing.
- Testing should be completed on worst-case product. What is the worst case product? How was the worst case product determined?
- What is the worst case sterilisation process?
- Is the labelling and final packaging reflective of the process/product going forward?

Stage 6- Packaging Validation

Medical Packaging Process Validation

The ultimate aim of process validation for a given packaging system is to demonstrate the manufacturing packaging process is fit for purpose and robust enough to meet the acceptance criteria as set out in product specifications.

Post Equipment Qualification (IQ- Installation Qualification/ OQ- Operational Qualification: Process validation which consists of Process Operational Qualification (OQ) and Performance Qualification (PQ) is required to be completed.

Operational Qualification: Operational Qualification challenges the worst case process sealing settings to ensure that worst case settings, product seal strengths and the other outputs meet specifications. OQ is typically completed for both high and low worst case conditions e.g. high temperature, high pressure and high seal times versus low temperature, low pressure or low time.

Performance Qualification: Performance Qualification provides a high degree of assurance that the sealing process will consistently produce a packaging system that meets predetermined specifications under normal operating conditions. Any product used in the PQ should be representative of the commercial process going forward.

A minimum of three lots is normally required for the PQ testing for an initial validation. Lots should be based on statistical rational and be reflective of commercial sizes.

Influencing Factors

Process validation aims to prove the consistency of a process under normal operating (anticipated) conditions. Every manufacturing process (however stable) is subject to influencing factors within day-to-day anticipated variation. These influencing factors can be categorised as (1) Worker, (2) Setup, and (3) Material.

For packaging equipment, these factors can influence the performance and consistency of a process and therefore need to be challenged during performance qualification.

A worker or operator can be a source of variation if the process is manual or not fully automated. Different people may have varying levels of expertise, experience and concentration. Even if an operator follows the instructions accurately and complies with standard operating procedures, there will likely be differences in how different operators handle raw materials and the product.

For packaging equipment that requires manual placement of raw materials, multiple operators should be used during the execution of validation builds.

Packaging systems are made up of more than one raw material or component. Whether components are manufactured in-house or supplied externally, they can be subject to variation, even if within the acceptance criteria. Therefore, for packaging validation, a minimum of three distinct lots of each material or component should be used. If multiple tooling is available, set-up can also be a source of variation. In these cases, a minimum number of setups should be completed as part of the validation. If the equipment is intended to run on different days or different shift patterns, the process may be subject to drift. Therefore, the validation should take account of this to ensure normal variation is captured.

Statistical Methods

The packaging process validation OQ and PQ must be performed using sample sizes and numbers of runs and batch sizes that are based on statistical rationale. A risk based approach should be taken when executing packaging validation.

Sample Size Determination and Sampling Rationale
The number of samples produced during the validation must meet predetermined acceptance criteria that are statistically relevant. This information should be documented in the pre-approved protocol.

Normality

For variable data (continuous data) a test for normality should be completed initially as this determines what type of statistical tool should be used. The normality of variable is verified by completing a normality test using a statistical package such as Minitab© or SPC© If the test returns a P-value ≥ 0.01, the data is considered normal. For non-normal data, accepted statistical methods for data transformation or non-parametric analysis should be used.

General Principles of Blister Packaging Validation

➢ The minimum and maximum critical process parameter set points at which product meeting all critical quality attributes can be manufactured should be defined during process development, prior to process validation.

➢ The calibrated range of all critical instruments must be greater than the process range (operating window).

> OQ-P testing will be executed by manufacturing worst case product at minimum process range settings (OQ-P min) and maximum process range settings (OQ-P max) for each critical process parameter. Devices manufactured at worst case minimum OQ-P min and OQ-P max settings must meet acceptance criteria for all critical quality attributes.

> Performance Qualification (PQ) should be manufactured at nominal process settings. Process capability (Cpk) or Process Performance (Ppk) critical quality attributes for each PQ batch.

Variable Print Packaging

This section examines some of the requirements for variable print which is either printed or laser etched onto the surfaces of labels, blisters or cartons.

What is Variable Print?

Variable print is when characters or other shapes or designs printed or laser etched on packaging materials need to change from lot-to-lot or product to product. Examples of variable print include lot/batch numbers, expiry date, date of manufacturing and other unique or variable information required to provide traceability and proper identification of the product.

What Is Fixed Print?

Fixed print is when characters or shapes do not change between lots or batches. The information (characters or shapes) is preprinted by the component supplier per approved artwork.

Packaging Definitions

Wicking: is the process in which through capillary action, moisture moves from the inside to the surface. During a dye penetration test, wicking can result in a false fail if the dye is exposed to the seal for too long.

Primary Packaging: The labelled inner container in which product is placed and sealed. This generally is the blister pack itself.

Secondary Package: The outer container into which one or more inner containers or "primary packages" are inserted into to form the complete finished product.

Ink Jet Print: applied by a printer that discharges liquid ink, one drop at a time, onto the component.

Laser Etching: "print" is applied by a laser that etches the characters or shapes into the material.

Print and Apply: Labelling method that prints variable print information on a label, then applies the label to a package.

Further Reading

- ASTM D1922: Test Method for Propagation Tear Resistance of Plastic Film and Thin Sheeting by Pendulum Method
- ASTM D1938: Test Method for Tear-Propagation Resistance (Trouser Tear) of Plastic Film and Thin Sheeting by a Single-Tear Method
- ASTM D1242: Resistance of Plastic Materials to Abrasions
- ASTM D3420: Pendulum Impact Resistance of Plastic Films

- ASTM F1306: Slow Rate Penetration Resistance of Flexible Barrier Films and Laminates
- ASTM D1709: Standard Test Method for Impact Resistance of Plastic Film by the Free Falling Dart Method
- ISO 10993:2009: Biological Evaluation of Medical Devices

CHAPTER 6
TEST METHOD VALIDATION

Introduction

This guidebook covers the design, execution and analysis of test method validation for medical devices. Test method validation involves the formal documentation of a test method used to capture and analyse data or information. The reason test methods need to be validated is to confirm that they are suitable and fit for the intended purpose. Secondly, and of equal importance is the need to verify that the test method performs to an acceptable level and is reliable and trustworthy over time. After all, test methods are used to assess product outputs such as dimensions, material strength and product functions. Getting this wrong will lead nowhere very quickly, so it is important to have confidence in the results of testing.

Validation studies must demonstrate method capabilities in the testing environment. As a result, validation studies allow the formal documentation of the ruggedness of the test method in real-use conditions (i.e. demonstrating that the precision and accuracy limits are met with different technicians, different production batches and variable test equipment, etc.).

Examples of Test Method Validations

Example 1

A packaging company has a seal strength on the lid of a package. It wants to put in place a test method to test the seal strength of the package. This scenario would call for a test method validation.

Example 2

A medical device incorporates the use of a spring that is used to actuate a valve. The manufacturer of the device wants to develop a test method that examines the tensile strength of the spring on an ongoing basis. This scenario would also call for a test method validation.

Example 3

A contact lens manufacturer uses an optical comparator to measure the diameter of contact lenses during manufacture. The manufacturer must develop and validate a test method to facilitate the measuring of contact lenses.

What is test method validation?

Test method validation is the documented process of ensuring a test method is suitable for its intended use.
The intended use of any system is normally documented and described in a User Requirements Specification. Test method validation involves establishing the performance characteristics and limitations of a method and the identification of influences which may change those characteristics.

Why should TMV be performed?

TMV is an important element of quality control. Without validation there can be no assurance that the test results will be reliable and fit for the purpose. In some fields, validation of methods is a regulatory requirement. Generally, any method used to produce data in support of regulatory (e.g., FDA) filings or the manufacture of devices for human use should be validated.

All are candidates for validation, though the process for each can vary. Most test methods exist as validated standards, methods developed by technical standard organizations (ANSI, ASTM, ISO…) to establish uniform methods and procedures for testing. But standard methods do not always fit the requirements of the tests to be performed.

When should methods be validated?

The risk associated with products an output (dimensions, features, chemical requirements etc.) often dictates the validation activity based on the potential level of patient harm weighed against the business risk of not performing the activities.

The device risk index or harm classification dictates the minimum level of statistical confidence required. Higher risk requires more rigorous testing and higher levels of statistical confidence. In most cases, TMV is not mandated in the medical device industry (except ISO 11607). But demonstrating the safety and effectiveness of a device is difficult to do if the methods for establishing these parameters are not shown to be appropriate and reliable. Test Method Validation may be required for:

- A new method is developed
- A revision to established methods
- Methods that are moved or transferred

Changes Requiring Re-Validation

Take the case where a standard test method is established and in operation. However, a change to the system software is required. This type of change could impact the measured output. Therefore the change needs to be considered for re-validation.

Any other changes to the test procedure such as a change in handling of test specimens or the change, addition, removal or modification of equipment including fixturing can impact the measured output.

It is important to note that validation of a test method is not required on each individual piece of equipment or fixturing, once replicate equipment or fixturing is assessed during the validation study.

Some examples of changes not necessarily requiring any re-validation or change to a validation report etc. include:

➤ Clerical corrections to the test method that do not change the method or affect the measurement of the output.

➤ Removing of referenced supplies that do not impact test output, for example lint or cleaning agents.

➤ Movement of equipment does normally not merit re-validation of the test method, but a limited equipment qualification may be required.

Test method validations should be product and site specific. This means the site and product should be clearly defined in the scope of the test method validation documents. Before an already existing validated test method can be used with a new product or at a new site, the suitability of the existing test method must be documented. Suitability reports are examined further along in this book.

Regulatory Overview

US Food and Drug Administration

The FDA defines a medical device as *"A medical device is an instrument, apparatus, implement, machine, contrivance, implant or other similar or related article, including a component part, or accessory which is;*

Recognized in the official National Formulary, or the United States Pharmacopoeia, or any supplement to them,

Intended for use in the cure, mitigation, treatment of disease, in man or other animals, or

Intended to affect the structure or any function of the body of man or other animals, and which does not achieve any of it's primary intended purposes through chemical action within or on the body of man or other animals and which is not dependent upon being metabolized for the achievement of any of its primary intended purposes."

The Code of Federal Regulations (CFR) Title 21 Part 820: Quality System Regulation (QSR) 21 does not clearly call out requirements on method validation. It does not actually mentions the words "method" and "validation" side by side. However, many warning letters have been issued to manufacturing on the subject since at least 2005. Method validation also protects the manufacturer from allowing defective product to circumvent inspection methods if not fit for purpose.

Regulating guidelines from a variety of sources covered in the sections below. The discussion, as it relates to method validation, is somewhat circuitous for medical devices. As stated previously, this is caused by the absence of the phrase method validation in FDA QSR documents. For this reason, some basic treatment relating to process validation is covered below even though this topic is covered in detail in a separate chapter especially because the actual CFR definitions are general enough to lump methods into the category of a process if a CDRH auditor sees fit to do so.

The first sentence of the 21 CFR 820 Quality System Regulation scope states:

*"Current good manufacturing practice (cGMP) requirements are set forth in this quality system regulation. The requirements in this part govern the **methods** used in, and the facilities and controls used for, the design, manufacture, packaging, labeling, storage, installation, and servicing of all finished devices intended for human use."*

In this first sentence, FDA has deemed the topic of methods as not excluded from the purview of FDA. This interpretation is evidenced by the warning letters and Form 483's issued to Medical Device companies described in later sections of this chapter.

FDA Title 21 Code of Federal Regulations Part 820.3

*(z) **Validation** means confirmation by examination and provision of objective evidence that the particular requirements for a specific intended use can be consistently fulfilled.*

(1) ***Process validation*** *means establishing by objective evidence that a process consistently produces a result or product meeting its predetermined specifications.*

(2) ***Design validation*** *means establishing by objective evidence that device specifications conform with user needs and intended use(s).*

(3) ***Verification*** *means confirmation by examination and provision of objective evidence that specified requirements have been fulfilled.*

(g) ***Design validation****. Each manufacturer shall establish and maintain procedures for validating the device design. Design validation shall be performed under defined operating conditions on initial production units, lots, or batches, or their equivalents. Design validation shall ensure that devices conform to defined user needs and intended uses and shall include testing of production units under actual or simulated use conditions. Design validation shall include software validation and risk analysis, where appropriate. The results of the design validation, including identification of the design, method(s), the date, and the individual(s) performing the validation, shall be documented in the DHF.*

(i) ***Design changes****. Each manufacturer shall establish and maintain procedures for the identification, documentation, validation or where appropriate verification, review, and approval of design changes before their implementation.*

(i) ***Automated processes****. When computers or automated data processing systems are used as part of production or the quality system, the manufacturer shall validate computer software for its intended use according to an established protocol. All software changes shall be validated before approval and issuance. These validation activities and results shall be documented.*

Title 21 Code of Federal Regulations Part 820.75

Sec. 820.75 Process validation

(a) Where the results of a process cannot be fully verified by subsequent inspection and test, the process shall be validated with a high degree of assurance and approved according to established procedures. The validation activities and results, including the date and signature of the individual(s) approving the validation and where appropriate the major equipment validated, shall be documented.

(b) Each manufacturer shall establish and maintain procedures for monitoring and control of process parameters for validated processes to ensure that the specified requirements continue to be met.

(1) Each manufacturer shall ensure that validated processes are performed by qualified individual(s).

(2) For validated processes, the monitoring and control methods and data, the date performed, and, where appropriate, the individual(s) performing the process or the major equipment used shall be documented.

(c) When changes or process deviations occur, the manufacturer shall review and evaluate the process and perform revalidation where appropriate. These activities shall be documented.

Title 21 Code of Federal Regulations Part 820.80

(a) General. Each manufacturer shall establish and maintain procedures for acceptance activities. Acceptance activities include inspections, tests, or other verification activities.

(b) Receiving acceptance activities. Each manufacturer shall establish and maintain procedures for acceptance of incoming product. Incoming product shall be inspected, tested, or otherwise verified as conforming to specified requirements. Acceptance or rejection shall be documented.

(c) In-process acceptance activities. Each manufacturer shall establish and maintain acceptance procedures, where appropriate, to ensure that specified requirements for in-process product are met. Such procedures shall ensure that in-process product is controlled until the required inspection and tests or other verification activities have been completed, or necessary approvals are received, and are documented.

<u>Title 21 Code of Federal Regulations Part 211.165 (e)</u>

(e) The accuracy, sensitivity, specificity, and reproducibility of test methods employed by the firm shall be established and documented. Such validation and documentation may be accomplished in accordance with 211.194(a)(2).

<u>Title 21 Code Federal Regulation Part 210.3 (b)</u>

(b) The following definitions of terms apply to this part and to parts 211, 225, and 226 of this chapter.

(1) Act means the Federal Food, Drug, and Cosmetic Act, as amended (21 U.S.C. 301et seq.).

(2) Batch means a specific quantity of a drug or other material that is intended to have uniform character and quality, within specified limits, and is produced according to a single manufacturing order during the same cycle of manufacture.

FDA requirements for medical devices, for validation apply to:

Processes—PPQs for processes including but not limited to manufacturing, sterilization, cleaning and packaging

Software—Any used in the device itself, manufacturing and test equipment, process control or quality processes

Equipment—IQ/OQ/PQs for Test Equipment and Manufacturing Equipment

FDA Medical Device Guidance Section 4: Process Validation

The Quality System (QS) regulation defines process validation as establishing by objective evidence that a process consistently produces a result or product meeting its predetermined specifications [820.3(z)(1)].

The requirement for process validation appears in section 820.75 of the Quality System (QS) regulation. The goal of a quality system is to consistently produce products that are fit for their intended use. Process validation is a key element in assuring that these principles and goals are met.

Processes are developed according to the design controls in 820.30 and validated according to 820.75. The process specifications, hereafter called parameters, are derived from the specifications for the device, component or other entity to be produced by the process.

The parameters are documented in the device master record per 820.30, 820.40 and 820.181. The process is developed such that the required parameters are achieved. To ensure that the output of the process will consistently meet the required parameters during routine production, the process is validated.

The basic principles for validation may be stated as follows:

- *Establish that the process equipment has the capability of operating within required parameters;*
- *Demonstrate that controlling, monitoring, and/or measuring equipment and instrumentation are capable of operating within the parameters prescribed for the process equipment;*
- *Perform replicate cycles (runs) representing the required operational range of the equipment to demonstrate that the processes have been operated within the prescribed parameters for the process and that the output or product consistently meets predetermined specifications for quality and function; and*
- *Monitor the validated process during routine operation. As needed, requalify and recertify the equipment.*

Per CDRH, the additional validation related definitions are:

Installation qualification: establishing documented evidence that process equipment and ancillary systems are capable of consistently operating within established limits and tolerances.

Process performance qualification: establishing documented evidence that the process is effective and reproducible.

Product performance qualification: establishing documented evidence through appropriate testing that the finished product produced by a specified process(es) meets all release requirements for functionality and safety.

Where process results cannot be fully verified during routine production by inspection and test, the process must be validated according to established procedures [820.75(a)]. When any of the conditions listed below exist, process validation is the only practical means for assuring that processes will consistently produce devices that meet their predetermined specifications:

Method validation as a requirement is not called out specifically; the FDA has issued warning letters and 483s in relation to the lack of "Method Validation".

W.H.O. Guidance

The World Health Organisation issued draft guidance for Test Method Validation of in vitro diagnostic medical devices in December 2016. Technical Guidance Series (TGS) for WHO:

Guidance on Test Method Validation of In Vitro diagnostic medical devices TGS-4

The guidance defines the terms verification and validation as follows:

*"**Verification** is the documentary proof that particular specifications have been met. When designing and developing an IVD, relevant attributes such as cost, and those for performance such as precision, sensitivity and stability are identified and given numerical specifications in design input documentation. It is subsequently the role of the R&D department to design an IVD that will meet those specifications.*

The R&D department consequently identifies valid test methods to demonstrate that the specifications have been met (verification) in the new design. Once design has been established, further numerical specifications are produced by the R&D department to ensure that the specifications of each attribute will be met consistently in routine production to ensure quality manufacturing. These new specifications are assigned to control critical production points and may include those for acceptance of raw materials, in-process materials, cleanliness of equipment, qualification of instrumentation and for the finalised IVD to verify its manufacture.

Again, it is also the role of the R&D department to identify appropriate test methods to monitor these specifications. An example of verification is related to incoming goods inspections; each time a raw material is purchased its properties will be verified against the specification using a validated test method."

Validation *is the documentary proof that the particular requirements for a specific intended use can be consistently fulfilled . Validation is defined as "verification against needs for a specific use" (i.e. the specification for that use). Within this guide, consistency is essential: it is an expectation that every lot of an IVD will behave as all other lots and will continue to meet design inputs. To ensure this, it is necessary to have validated test methods for measuring and/or monitoring specifications that will consistently produce results fit for purpose. The test methods must be validated to ensure that the results of measuring and/or monitoring are meaningful. For example, the need for accurate measurement of a raw material weighed in micrograms will not be achieved by using a weighing device with tolerance measured in grams. A test method using such an instrument would not be valid for the intended use. Thus, for the example provided, a test method should be specified that has the necessary accuracy and precision for measuring such weights, and an instrument and procedure identified that will consistently achieve this requirement during its use. The test method is then validated to produce results fit for purpose.*

Validation of a test method is distinct from its characterisation. Characterisation is documentation of some or all of the features of the method; validation is ensuring that the relevant characteristics are appropriate for the specific intended use. Validation of a method to be used widely, and for standard methods, often begins with complete characterisation. However, for each specific intended use it is likely that only a subset of the characteristics will be relevant and must be evaluated.

ISO 13485, Medical Devices Standard

Introduction

ISO 13485 is the quality management standard of choice for manufactures of medical devices. Revised in 2016, ISO 13485:2016 "specifies requirements for a quality management system where an organisation needs to demonstrate its ability to provide medical devices and related services that consistently meet customer and applicable regulatory requirements."1 The scope of the standard can apply to any organisation or company involved throughout the life-cycle of a product, including design and/or development, production, storage and distribution, installation, or servicing of a medical device and design and development or provision of technical or professional services.

The recent revision is designed to address recent developments in quality management and other updated regulations that relate to the industry. Improvements in the new version of the standard include broadening its applicability to include all organisations involved in the life cycle of the product, from the concept stage to end of life along with greater alignment with regulatory requirements and post-market surveillance including complaint handling.

ISO 13485:2016 is also used by suppliers or external vendors that provide QMS related management system- services. Requirements of ISO 13485:2016 are applicable to organisations regardless of their size and regardless of their type except where explicitly stated. Wherever requirements are specified as applying to medical devices, the requirements apply equally to associated services as supplied by the organisation. If any requirement in Clauses 6, 7 or 8 of ISO 13485:2016 is not applicable due to the activities undertaken by the organisation or the nature of the medical device for which the quality management system is applied, the organisation does not need to include such a requirement in its quality management system. For any clause that is determined to be not applicable, the organisation records the justification as part of their certification and quality management system.

Basic Definitions as defined in EU Annex IX of Directive 93/42/EEC)

Intended Purpose: Intended purpose means the use for which the device is intended according to the data supplied by the manufacturer on the labelling, in the instructions and/or in promotional materials. (Chapter I section 1 of Annex IX of Directive 93/42/EEC)

Transient: Normally intended for continuous use for less than 60 minutes.

Short Term: Normally intended for continuous use for not more than 30 days.

Long Term : Normally intended for continuous use for more than 30 days.

Invasive Devices: A device which, in whole or in part, penetrates inside the body, either through a body orifice or through the surface of the body.

Body Orifice: Any natural opening in the body, as well as the external surface of the eyeball, or any permanent artificial opening, such as a stoma.

Surgically Invasive Device: An invasive device which penetrates inside the body through the surface of the body, with the aid of or in the context of a surgical operation.

Implantable Device: Any device which is intended:

- to be totally introduced into the human body or,
- to replace an epithelial surface or the surface of the eye, by surgical intervention which is intended to remain in place after the procedure. Any device intended to be partially introduced into the human body through surgical intervention and intended to remain in place after the procedure for at least 30 days is also considered an implantable device.

Medical Device: means any instrument, apparatus, appliance, material or other article, whether used alone or in combination, together with any accessories or software for its proper functioning, intended by the manufacturer to be used for human beings in the:

- diagnosis, prevention, monitoring, treatment or alleviation of disease or injury.

- investigation, replacement or modification of the anatomy or of a physiological process.

- control of conception which does not achieve its principal intended action by pharmacological, chemical, immunological or metabolic means.

A medical device may be assisted in its function by the following means:

Active Medical Device: any medical device relying for its functioning on a source of electrical energy or any source of power other than that directly generated by the human body or gravity.

Active Implantable Medical Device: any active medical device which is intended to be totally or partially introduced, surgically or medically, into the human body or by medical intervention into a natural orifice, and which is intended to remain after the procedure.

Custom-Made Device: means any active implantable medical device specifically made in accordance with a medical specialist's written prescription which gives, under his responsibility, specific design characteristics and is intended to be used only for an individual named patient.

Device Intended for Clinical Investigation: any active implantable medical device intended for use by a specialist doctor when conducting investigations in an adequate human clinical environment.

Intended Purpose: means the use for which the medical device is intended and for which it is suited according to the data supplied by the manufacturer in the instructions.

Putting into Service: means making available to the medical profession for implantation.

Where an active implantable medical device is intended to administer a substance defined as a medicinal product within the meaning of Council Directive 65/65/EEC of 26 January 1965 on the approximation of provisions laid down by law, regulation or administrative action relating to proprietary medicinal products (6), as last amended by Directive 87/21/EEC (7), that substance shall be subject to the system of marketing authorisation provided for in that directive.

Where an active implantable medical device incorporates, as an integral part, a substance which, if used separately, may be considered to be a medicinal product within the meaning of Article 1 of Directive 65/65/EEC, that device must be evaluated and authorised in accordance with the provisions of this directive.

ISO 13485 & Regulations

In Europe, EN ISO 13485:2013 helps companies meet the requirements of: Directive 93/42/EEC on medical devices. This harmonised standard gives companies the "presumption of conformity" to complying with directives.

EN ISO 13485 was published in February 2013 and harmonised in August 2013 to cover the three directives:

- 90/385/ECC– The Active Implantable Medical Devices Directive (AIMDI)
- 93/42/ECC – The Medical Devices Directive (MDD)
- 98/79/EEC – In Vitro Diagnostic MDD (IVDMDD)

Overview of Standard

ISO 13485 has 8 Clauses or Sections which make up the structure of the standard.

Section 0 Normative References, Definitions and Terms

Section 1 Requirements of the Quality Management System (QMS)

Section 2 Normative References

Section 3 Terms and Definitions

Section 4 Requirements of the Quality Management System (QMS)

Section 5 Management Responsibility

Section 6 Resource Management

Section 7 Product Realisation

Section 8 Measurement, Analysis and Improvement

With regard to Test Method Validation, the relevant areas of ISO 13485 include:

(1) Clause 7: Product Realisation- Section 7.3 Design and Development

(2) Clause 8: Measurement Analysis

Clause 7: Product Realisation- Section 7.3 Design and Development:

Design and Development Verification and Validation ensure that the product is designed, developed and subsequently manufactured meeting all the customer requirements, regulatory requirements and business requirements. These requirements are classed as inputs to the design and development, and verification and validation ensure the inputs have been adequately taken into account.

The design and development testing sometimes replicate the commercial applications of the medical device, hence providing a realistic challenge in order to have confidence in the medical device.

Design Control

Design control is a necessary practice that ensures good engineering principles are maintained throughout the design phase of a product. It also refers to the continual design and development of the product through its very lifecycle. The design and development files and history must be controlled and maintained, with any changes properly assessed, tested and documented.

Clause 8: Measurement Analysis:

Clause 8 includes:

8.1 General requirements
8.2 Monitoring and measurement
8.3 Control of nonconforming products
8.4 Analysis of data
8.5 Improvement

8.1 General Requirements

Measurement, analysis and improvement are the key themes of clause 8. As with all medical devices, inspection and testing both during manufacturing and post manufacturing is necessary to ensure products and services function as intended and without defects. With any type of measurement or inspection analysis, the method used to complete the testing is critical. The method must be fit for purpose, and the equipment must be suitable. This "method validation" typically is done during the design and development phase.

8.2 Monitoring and Measurement

Monitoring and measurement are dependent on the information or feedback provided from various sources. The most important feedback is the post-production feedback that is gathered from customers or the end user. Again, this occurs over the whole lifetime of the product or service in question. There are a number of methods that can be used to obtain feedback. Some examples include:

- Customer surveys
- Customer complaints
- Review of regulatory databases such as MAUDE.
- Repair and servicing information

8.3 Control of Nonconforming Product

Non-conforming product presents a risk to patients or users of medical devices. When a situation arises where non-conforming product is manufactured or detected through inspection processes, the product must be controlled and segregated to prevent unintended use or distribution.
Some examples resulting in non-conformance are:

- When a manufacturing process drifts outside its validation window or operating parameters.
- A certificate of analysis for a raw material is not provided by the supplier or the results are out of specification.
- In-process testing was not completed at the defined intervals.
- Training of personnel completing tests is not current or is inadequate.

8.4 Analysis of Data

In any engineering activity, data and the quality of the data is a key factor in making the right decisions. Provided the data collected is relevant and accurate, analysis of data can provide important insights into process performance, quality control and product functionality. Data should be collated in a consistent way and controlled by a procedure. When it comes to medical device manufacturing, the sources and types of data are multiple. Data can be generated from in-process testing and data can be generated from end of line testing aka finished product testing.

8.5 Improvement

ISO 13485 fosters a culture of continual improvement. As we have seen, each activity can be described as a process. For example, a manufacturing process, a procurement process, a complaints process. The set of processes that make up the quality management system need to be continually reviewed to ensure they are suitable and effective for the task at hand. Typical tools used to keep improvement in mind include:

- Review of the quality policy and quality objectives
- Frequency and category of corrective and preventative actions (CAPA's)
- Customer complaints

- Management review input

Definitions and Key Concepts

Attribute: is defined as the result of a property or characteristic. It is generally used with the terms pass or fail.

Accuracy: can also be defined as trueness. An expression of the closeness of agreement between the value that is accepted, either as a conventional true value or an accepted reference value and the value obtained. A system with low bias implies good accuracy and vice versa.

ANOVA (Analysis of Variance): a statistical method used to evaluate the significance of differences in means due to different factor-level combinations.

Bias: The difference between observed "average of measurements" and a reference value; also referred to as accuracy.

CQA (Critical-to-Quality): a property or characteristic with specific nominal value and appropriate limit and range providing a particular quality attribute.

Critical Process Parameter (CPP): a process parameter that has a direct impact on critical quality attributes.

Dichotomous Variable: an output with only two possible values. Also known as dummy or indicator variable.

Equipment Qualification: establishing documented evidence that the process equipment is suitable for the intended use and is capable of consistently operating within established limits and tolerances under normal operating conditions.

Process Validation: process validation is defined as confirmation via documented evidence that a particular process performs consistently to a high degree of assurance in accordance with predetermined specifications under anticipated conditions.

Measurement Capability Index (MCI): the Measurement Capability Index (MCI) represents the capability of the measurement system. It is used to evaluate the capability of the gauge to classify product against predetermined specifications.

MSA: a study to determine the degree of error involved in measuring the given parameter. The measurement system involves the combination of operations, procedures, gauges, instruments, environmental conditions, people and software.

Precision: the degree of agreement (scatter) between a series of measurements when a method is applied repeatedly to multiple samplings of a homogeneous sample or artificially prepared sample under the prescribed conditions. There are three types of precision; repeatability, intermediate precision and reproducibility.

Range: range is defined as the interval between the upper and lower measurements required. The minimum specified range should be within the equipment range and validated to operate at all points within the range.

Ruggedness (Intermediate Precision): variation on different days or with different analysts and equipment. The extent to which intermediate precision should be established depends on the circumstances under which the method is intended to be used.

Resolution: the smallest unit of measure that can be obtained reliably from a measurement device, also known as gauge discrimination.

Gauge R&R: represents the estimate of the measurement variation. The measurement variation has two components; repeatability or the precision under the same operating

140

conditions (same operator, test method, sample, etc.) and reproducibility or the precision between operators when measuring the same sample with the same gauge.

Variable: is generally the output that is measured.

Validation: confirmation by examination and provision of objective evidence that the particular requirements for a specific intended use can be consistently fulfilled.

New Test Methods

A test method procedure should be created as early on as possible and trialed and examined for completeness and appropriateness.
If new test methods are required, a revision controlled draft should be available for the purposes of the test method validation.

Changes to Existing Methods

If changes to existing test methods are required, a redlined version highlighting the changes should be made available for the test method validation.

Method Transfer

If an existing test method is suitable for the test method validation, a suitability report can be completed to document the suitability and show that all factors have been considered (see attachment 1). However, the test method should have been previously validated. The parameters at which the validation is to be conducted must be within the existing validated range.

Equipment used in a test method must be assessed to ensure the process is within the equipment qualification. All validation testing must be done on qualified equipment. Equipment qualification is therefore a prerequisite of test method validation.

Test Method Ruggedness Study Protocols

Ruggedness refers to the variation, on different days or with different operators or equipment. The extent to which ruggedness (aka intermediate precision) should be established depends on the circumstances under which the method is intended to be used.

An initial ruggedness assessment should be completed to understand the sources of variation. More formal ruggedness studies may be required which should be captured in a formal study protocol.

The output of any ruggedness studies should detail any changes or modifications to the test method procedure.

Generally, a scoring system is used to describe ruggedness which forms a ruggedness assessment. As a result of ruggedness studies and consequent updates to the procedure, the ruggedness assessment needs to be reassessed. This reassessment should be reflected in the final scores of a Ruggedness Assessment Matrix.

Accuracy

Accuracy is a measure of exactness of the test method output or another way of putting it is the closeness of agreement between a set of test results.

For example, take a component that weighs exactly 4 kg according to an NIST traceable scale. If the weight of component is taken 10 times on the balance under study using the test method under study then calculate the mean weight of the 10 readings. The offset between the mean weight and the 4kg "accepted reference value" is a measure of bias.

A large bias = poor accuracy. A small bias = good accuracy.
It is important to note that accuracy does not address the variation between individual measurements.

Simply put, if the average is very close to 4kg, then the test method could have been declared to be very accurate.

It is advised that you consult any relevant standards (e.g. ISO, ASTM) to the product or feature being measured as standards often will call out an accuracy requirement. Generally, results should be accurate to $\pm 1\%$ of the measured value. Therefore, the equipment must be fit for the intended purpose or the measurements in mind.

Note: instrument or equipment accuracy can normally be found on calibration certs provided by the manufacturer or vendor.

☐

Precision

The precision of a method is the degree of agreement among individual test results when the same test method or procedure is applied repeatedly to multiple samplings that represent a population.

Precision can be a measure of either the degree of reproducibility or of repeatability of the method.
Repeatability refers to the use of a method using the same operator/test person with the same equipment. Repeatability should be assessed using either a minimum of 9 determinations covering the specified range for the method (e.g. 3 concentrations /3 replicates each). Reproducibility refers to the use of the analytical method in different laboratories such as in a collaborative study.

Ruggedness

Intermediate precision (also known as ruggedness) expresses differences related to laboratory variation, as on different days, or with different analysts or equipment within the same laboratory. The extent to which intermediate precision should be established depends on the circumstances under which the method is intended to be used. The effects of random events on the precision of the analytical method should be established. The use of experimental design (matrix) may be used to study the effects of typical variation (dominance factors) on the analytical method (e.g. equipment, analyst, days).

Representative/Continuous Sampling

Representative sampling is used to determine overall process performance (e.g. Pp / Ppk), which is more applicable for processes known or suspected as less than stable or not in statistical control. Sampling in this way best determines overall spread, which includes within-time and time-to-time variation.

Below, some examples are given on how to sample representatively:

1. Sampling over a given time-period: e.g. a tray of product is produced every 15 minutes, the period of interest is a 1 hour interval and the sample size is 40.

2. Sampling a batch or product lot not assembled in any order: if the product is packed in a tray (without any grouping) then sample from various sections of the tray.

<u>Consecutive Sampling</u>

This type of sampling involves taking one sample immediately after each another for the subgroup or time period in question, and is used to determine process capability (e.g. Cp / Cpk).

Consecutive sampling is used in particular to create control charts where a process is sampled in time order by selecting a subgroup sample consecutively and repeating this sampling over a number of subgroups while in same time order.
This method is typically used when the process is stable as there will be little or no causes of lot-to-lot variation.

□

Range

The range is defined as the interval between the upper and lower measurements required. The minimum specified range should be within the equipment range and validated to operate at all points within the range.

If an existing test method or piece of equipment is to be used, it is important to determine if the method parameters for the new/modified test method are within the validated range of the equipment qualification. Remember, all validation testing must be done on qualified equipment. Typically, the equipment qualification assessment is documented in the test method validation protocol.

Resolution

We have previously defined resolution as the smallest unit of measure that can be obtained reliably from a measurement device or system.

For example, a Vernier callipers may have different models with different resolutions. Some will have only two digits to the right of the decimal point (X.XX mm) and other models could read three digits to the right of the decimal point (X.XXX mm).

The instrument resolution should be better than the resolution of the product specification. If the product specification is X.XXX, then at least a "four-digit" measurement device should be used.

☐
Probability Of False Alarms P (Fa)

This signifies the likelihood of rejecting a conforming unit. This is typically an acceptance criterion for attribute tests. Refer to MSA template for further illustration.

Probability Of Misses P (M)

This indicates the likelihood of accepting a non-conforming unit. This also is typically an acceptance criterion for attribute tests. Refer to MSA template for further illustration.
☐

Validation Protocols

Typically, an approved template is used to create a validation protocol. The protocol sets out the approach to the validation i.e. the approach to qualify the test method. Refer to the appendix for an example of a validation protocol template.

Attachments to the protocol should include ruggedness assessments completed and references to supporting studies/reports. The drafted or "redlined" test method should be attached to the protocol also. The type of MSA protocol (attribute or variable) should also be determined in the validation protocol.

☐

What Can Impact the Accuracy of a Test Method?

Accuracy is influenced by both the instrument (scale) and the test method. If you drop the object on the scale and take a reading before the scale has stabilised, the accuracy is likely to be poorer than when using a test method that demands allowing the scale to stabilise.

Examples include: Tensile strength at break - strength does not exist as a material property independent of the test method used to measure it.

For properties like time, distance, and mass, there are NIST traceable standards that can be measured. These standards have a generally accepted reference value that can be compared to the observed readings to assess accuracy (bias). No such reference sample exists for tensile strength at break, deflation

time or implant radial strength. For tests without a reference value, the accuracy of the underlying sensor (e.g. load cell) used to determine the output should be addressed if possible.
☐

MSA Studies

A measurement system analysis (MSA) is an experimental design used to identify the elements that affect measurement variation. There are two types of data in which MSA studies can be completed i.e. variable data and attribute data. These terms are defined below. Variable data: data that can assume a range of numerical responses on a continuous scale. Most measurements yield variable data.

Attribute data: data that represents the absence or presence of a characteristic.

Non-destructive tests: test where the measured characteristic is not altered due to testing. Since the sample is not altered, multiple readings can be taken on the sample with the expectation of getting the same measured result.

Destructive tests: test where the measured characteristic is changed due to testing. Since the sample is changed, there is no expectation of getting the same measured result over multiple readings.

So, in summary that makes up four types of MSA studies:

- ➢ Variable / Non-Destructive
- ➢ Variable / Destructive
- ➢ Attribute / Non-Destructive
- ➢ Attribute / Destructive Table

The following sections describe the requirements, measurement capability indexes and the typical acceptance criteria per MSA type.

General MSA Requirements

Test Environment Conditions - the test environment (i.e. temperature, humidity) should represent the conditions going forward. The effect of multiple environmental conditions can be evaluated if the study is properly designed and planned.

Sample Range - samples should cover the expected range of measurements.

Standard (for attribute MSA) - define the true answer (pass or fail). The standard is based on the inspection ratings of an expert opinion or a measurement system with known better inspection capability than the one under evaluation.

Measurement Instructions/ Training - follow the inspection instructions as defined in the controlled documents or redlines included with the protocol. Do not minimise variability by adding special instructions not defined in the controlled documents or redlines included with the protocol. Reference the controlled documents in the protocol. Special instructions are allowed when using pseudo samples provided that the variability is not minimised due to the instructions. Testers should have a high degree of skill and experience. Do not use new personnel or inexperienced people to conduct measurement studies.

Equipment Qualification and Calibration – The equipment must be calibrated prior to conducting the study. Evidence of the calibrated state should be documented in the report (e.g. calibration certificates etc.). It is important not to re-calibrate the equipment during the study as results can be different due to the calibration effect. The effect of calibration can only be evaluated if the study is properly designed.

Randomisation –

1. Assign the samples to the first operator in random order. Operator measures the parts.

2. Assign the samples to the second operator in random order. Operator measures the parts.

3. Assign the samples to the third operator in random order. Operator measures the parts. Repeat the process described in steps 1 to 3 with the operators for a second and third trial.

Data collection - when documenting the results of a trial, the operator should not have access to the results from the previous trials. A different data collection sheet must be provided for each operator involved in each trial. In lieu of a different data sheet, a data recorder may be used to blind the data recording operator to the test data of previous runs.

Variable MSA Studies

Non Destructive/Variable Msa Studies

The key requirements for non-destructive and variable MSA studies include:

No. of Operators – at a minimum, 3 operators should be used during the study. More operators are also recommended if human/operator interaction is a source of measurement error.

Sample Size – a minimum of 10 units is recommended.

Trials - a minimum of 3 trials should be completed.

Destructive/Variable Msa Studies

If a test is destructive in nature, repeated measurements cannot be taken as the sample is damaged or destroyed as part of the test. One solution is to adopt standardisation of units where homogeneous samples are created by standardising the material or manufacturing process.

- No. of Operators – 3
- Sample Size – 10 units
- Trials – 3 trials

This equates to 90 measurements in total. If standardisation is not feasible, the use of non-destructive pseudo-samples can be used. However, equivalence should be demonstrated between the pseudo sample and "true" units.

Attribute MSA Studies

Non destructive

The recommended and minimum sample size requirements for attribute/non-destructive MSA studies are shown below:

Recommended Minimum Sample Size Requirement

- # Operators - 3
- # Sample size - 25
- # Trials – 3

Destructive

When the test is destructive, repeated measurements cannot be taken as the sample is destroyed or altered. Some approaches are outlined below in order to quantify the measurement variability for destructive tests.

Standardisation Approach: homogeneous and representative samples are created by standardising the method of sample preparation, or material.

Sub-samples: cut each sample into three sub-samples to represent the three trials.

Pseudo-samples: create non-destructive pseudo-samples, documenting a rationale justifying the equivalence of the pseudo samples to the true samples.

Measurement Capability Index

The Measurement Capability Index (MCI) is calculated to assess the capability of the measurement system. The MCI is calculated as a % tolerance.

Measurement Capability Index acceptance criteria:

This index is used to evaluate the capability of the gauge to classify product against the specifications.

The index represents the % of the tolerance (upper specification limit (USL) and the lower specification limit (LSL) that is consumed by the measurement system variation. Figure 9 shows a graphical representation for this index.

Action Plan - Identifying Sources of Variation

Sources	Action
Measurement variation due to repeatability	Ensure the sample is not deformed over time due to repeated measurements and trails
	Ensure the equipment specification has the required precision
Measurement variation due to reproducibility	Review training and introduce standard work instructions

☐

Suitability for Use Rationalisation Report Template

If an existing test method can be used with no or minor changes, a Test Method SFU Report can be used to document the test method validation.

Suitability for use report is appropriate only if the new product test method parameters are within the existing validated range.

If the test method parameters for the new product are outside of the validated range, the test method must be re-validated. Examples of cases which can utilise such suitability for use reports include:

> ➤ Test method transfer to a new manufacturing site.
> ➤ New product where the product specifications fall within the validated output range.

➢ Minor changes in component material which do not impact the validated test. Examples of changes that require full validation include:

New products:

➢ Extension of product sizes that fall outside the validated range.

☐

Appendix 1

Suitability for Use Rationalisation Report Template

Test Method Suitability for Use Rationalisation Report

Product Name

Site

Author:

Date:

Approvals:

1. PURPOSE

The purpose of this report is to document the justification for use of XYZ assuming that XYZ has been previously qualified and validated for product XYZ.

The conclusions of the test method validation/suitability for use are valid and can be leveraged for new product zzz to be tested at Site YYY.

2. SUMMARY

This report documents evidence of the suitability for use of the test method XZY.

Table 1 - Variable Data

Test Method	Measurement	Precision	Validated Range	Resolution
TMXXX	Tensile Strength	MCI = 16%	0% to 11%	0.01 %

Table 2 - Attribute

Test Method	Measurement	Effectiveness	P(FA)	P(M)
TM000	Laser Mark	99%	1%	1%

4. CONFORMANCE TO GUIDELINES AND STANDARDS

Reference	Title	Revision
EN ISOxyz	*Xxyyzz*	*Date*
FDA	*Xxyyzz*	*Date*

4. JUSTIFICATION

The following are the justifications for leveraging test method suitability for use Justification for Leveraging xxxyyyzzzz validation conclusions for new product xxyyzz at site xxyyzz.

Test Method	Leveraged Report(s)	Parameter	Leveraging Justification
X	XXYYZZ	Probability of False Alarms P(FA)	

Test Method	Leveraged Report(s)	Parameter	Leveraging Justification
		Probability of Misses P(M)	
		Effectiveness	
		Ruggedness	
Variable Outputs (Continuous Data)			

Test Method	Leveraged Report(s)	Parameter	Leveraging Justification
X	Xxxyyyzzz	Accuracy	The acceptance criteria for accuracy may be driven by a sandard such as ISO, ASTM. Generally, the calibration documentation will detail the accuracy of test equipment. If the equipment is capable of measuring to 1% of the measured value, the system can be deemed accurate.

Test Method	Leveraged Report(s)	Parameter	Leveraging Justification
Y		Precision	MCI ≤30% is typically the acceptance criterion. MCI is calculated via data captured during an MSA (Measurement System Analysis). Using a statistical tool the MCI is calculated.
		Range	The test method range and the equipment range must both meet the requirements of the product feature to be measured.
		Resolution	Resolution is the smallest unit of measure that can be obtained reliably from a measurement device, also known as gauge discrimination. E.g. 0.01mm.

Appendix 2

Test Method Validation Protocol

Test Method Validation Protocol

TEST NAME
PRODUCT/SITE

Author:

Date:

Approvals:

1. PURPOSE

The purpose of this validation protocol is to document the requirements and acceptance criteria that will establish that test method XYZ is suitable for use with product X at site Y for testing the characteristics listed in Table 1.

Measurement	Specification
TEST 1-Tensile Strength	XXXYYYYY YYZZZZZZZ
TEST 2- Dimensional Length	XXXXXXYY YYYYYYYY YYYYZZZZZ ZZZZZZZ

2. SCOPE

The requirements and acceptance criteria as specified in the test method requirements.

3. DEFINITIONS

For the purpose of this validation protocol, the following terms and definitions apply:

RPT *Report (document prefix)*
TM *Test Method (document prefix)*

5. REFERENCES

Reference	Title	Revision
EN ISO XXXX	XYZ	DATE
FDA Guidance	XYZ	DATE

6. BACKGROUND

Verification of Equipment Qualified Ranges

Equipment	Parameter	Qualified Range of Equipment	Test Method Range	Pass
XYZ	*Diameter*	*E.g. 1 to 20mm*	*2.0mm – 15.0mm*	*Yes*

7. CONFORMANCE TO GUIDELINES AND STANDARDS

Guideline/Standard	Measurement	Requirement
EN ISO XYZ		
FDA XYZ		

8. ACCEPTANCE CRITERIA AND RATIONALE

Parameter	Requirement	Acceptance Criteria	Verification Process
Accuracy	Reference accuracy requirements specified in standards. E.g. accuracy of ±2%	±2°C	Review the instrument measurement accuracy by referencing calibration records

Precision	MCI.	MCI ≤ 30%	A Measurement System Accuracy GR&R Study to be completed.
Range	Define the range of measurements	Test method must be able to measure values within the required range.	Parameters and settings should be justified
Resolution	0.1mm	0.1 °C	The resolution of gauge or equipment as per calibration assessment.

CHAPTER 7
MEASUREMENT

Introduction

This chapter covers the basic principles of linear measurement. This involves the comparison of the piece under test with a known standard. These days, many manufacturing processes utilise automated measurement or complex vision systems. However, an understanding of the basic principles of measurement along with traditional measurement methods is fundamental to the engineer. Many of these traditional measurement methods are contact in nature (opposed to non-contact). While modern vision systems tend to be non-contact in nature, which offers multiple benefits.

Historically, there were two systems in general use: the imperial or English system and the metric system, with the latter a clear preference for the majority of In the search for an absolute standard, lightwaves as a standard of length are used for measuring in many applications. The lightwave method has the advantage of being non-variable, repeatable and of very accuracy. With the absolute standard defined in terms of lightwaves, a technique known as interferometry is used to measure and calibrate many master standards.

Working standards are calibrated against master standards meaning that a comparison between the absolute standard and the particular measuring instrument is achieved. The calibration and testing of all measuring tools and instruments usually requires special equipment and should be completed in accordance with a recognised calibration standard.

Non-contact methods of measurement do not require physical contact with the part under measurement. The measurement is achieved by the use of optics, lasers or magnification. This can be an advantage as the part does not get exposed to physical forces during measurement. Non-contact methods also eliminates the "feel factor" and resulting human error of hand gages.

Contact measuring methods include micrometers and plug gages where accuracy and consistency are subject to the amount of pressure applied to the part during measurement.

Care & Maintenance of Instruments

1. Keep all instruments clean, treat them with care and avoid misuse.
2. Place instruments in cases or fit covers when not in use.
3. Keep the inside of the instrument case clean. The case is meant to protect the instrument.
4. Do not attempt to dismantle an instrument. If it is not functioning correctly return it to the appropriate department for servicing.
5. Choose the instrument in keeping with the tolerance of the dimension to be measured.
6. Wherever possible use an instrument that gives a direct reading.
7. Do not use worn or damaged instruments.
8. Remember that the graduations on a measuring instrument (resolution) are not necessarily the accuracy to which it can be used. Quite often the in-built inaccuracies of measuring instruments exceed their resolution.

Manual Measuring Instruments

The ease with which a component can be measured and the accuracy to which it need be or can be measured depends on the correct choice of measuring instrument. Factors such as the shape of the component and the position of the dimension to be measured influence the choice of instrument but nevertheless, certain basic rules should be followed.

Instrument or Measurement	Type of Measurement	Value of Smallest Graduation (Resolution)	Suggested Reliability/ Accuracy
Steel rule	direct	.5 mm.	±.5 mm.
Depth gauge	direct	.5 mm.	±.5 mm.
Calipers	direct	none	±.5 mm.
Vernier calipers	direct	.01 mm.	±.05 mm.
Vernier depth gauge	direct	.01 mm.	±.05 mm.
Vernier height gauge	direct	.01 mm.	±.05 mm.

Micrometers			
Instrument or Measurement	Type of Measurement	Value of Smallest Graduation (Resolution)	Suggested Reliability/ Accuracy
25-50 mm.	direct	.01 mm.	±.01 mm.
150-300 mm. plain	direct	.01 mm.	±.01 mm.
150-300 mm. plain	direct	.01 mm.	±.02 mm.
Inside micrometres	direct	.01 mm.	±.01 mm.
Depth micrometre	direct	.01 mm.	±.01 mm.
Telescopic gauges	transfer	none	±.02 mm.
Slip gauges	end standard	.001 mm.	±.0005 mm.
Dial test indicator	comparison	.01 mm.	±.01 mm.
Dial test indicator	comparison	.001 mm.	±.001 mm.

The volume of measurements is also a limiting factor. While a properly trained individual may be able to make repeated measurements, automation should be considered for repetitive inspection if possible.

Slip Gauges

Slip gauges or block gauges are used as standards for precision length measurement throughout the engineering industry. The gauges are usually made in sets and consist of a number of hardened steel blocks. Each block has two of the opposite faces lapped flat and parallel to a definite size within an extremely tight tolerance. In building up packs of slip gauges, errors can occur if care is not exercised. There are three main causes of errors:

1. Errors due to deviation from true size.
2. Errors due to grease or dirt between the wringing faces.
3. Errors due to expansion caused by excessive handling or leaving gauges exposed to strong sunlight or electric lamps.

The Sine Bar

The sine bar, which is one of the most effective methods of precision measurement of angles consists of a rectangular section bar to which are attached two hardened rollers of the same diameter, such that the common centre line of the rollers is parallel to the top face of the bar. The principle of the sine bar is based on the fact that in a right angled triangle the function known as the sine of the angles is the relationship of the hypotenuse to the side opposite the angle.

$\operatorname{Sin} A = a/c$

To set up a sine bar to a required angle:

(a) Select and clean the appropriate bar.
(b) Determine the sine of the angle required by reference to the sine tables.
(c) Calculate the slip gauges required, i.e. multiply the sine of the angle by the size of the sine bar, i.e. the distance between the centres of the rollers.

(d) Select and clean the necessary slip gauges including protector slips.
(e) Wring the slip gauges together.
(f) Clean the surface plate and set up the bar by inserting the slip gauges under the roller as shown.

Other Applications of the Sine Principle

Sine centres are a very useful adaption of the sine bar principle. Adjustable centres are mounted on the face of the sine bar allowing such items as taper gauges and taper shafts to be inserted between centres and be measured for both angle and concentricity.

Vision Measuring Systems

While manual measurement via the previous instrumentation is useful in certain circumstances of low volume manufacturing, the requirements of high volume manufacturing and reliable quality calls for automated vision measuring systems to meet the volumes of inspection required.

Coordinate Measuring Machines

Coordinate measuring machines are available in a wide range of sizes and accuracy and can meet most precision 3D measuring applications for today's needs.

Depending on the system setup and capability, a wide range of contact and non-contact probes can allow numerous kinds of measurements to be performed. CMM software assists in the analysis and interpretation of measurement results, which is particularly useful with increasing quantities of measurement.

Optical Comparators

Often referred to simply as a comparator, they are used to measure parts dimensionally by using optics and projection. Measurement is achieved by overlaying limits or graduations over the image projected. Inspection with comparators is relatively quick and is most useful when looking for a pass/fail result.

However, the rise of vision-based inspection systems makes the manual comparator, even equipped with modern capabilities, an oft overlooked tool. This is particularly the case when the requirement is to inspect large quantities of parts at once, since a vision system allows you to place multiple parts for inspection on the stage at the same time. For many simple measurements on two-dimensional parts with clearly defined edges, the optical comparator is quite suited.

Optical comparators can provide more information than just simple dimensions. Length and width measurements of the part shown above, for example, can be quickly obtained from two separate measurements by using a micrometer. These superficial measurements, however, might not reveal burrs, scratches, indentations or undesirable chamfers. Such imperfections are best detected on a comparator. In addition, a comparator's screen can be simultaneously viewed by more than one person and provide a medium for discussion, just as a white board might facilitate a conference.

Advantages

- Fast length and width measurement
- Length and width measurement can be done simultaneously
- Burrs and chamfers can be detected
- Screen can be viewed by more than one person

Digital Comparators

Digital comparators allow Pass/Fail inspection comparisons, Files are simply uploaded to the equipment in order to compare measurement results against a CAD overlay - the overlay moves with the datum, so you don't even need to use an optical comparator / profile projector.

Geometric Tolerancing

Before an object measured, complete information about both the size and dimensions and tolerances of the object must be available. The shape of an object is communicated through orthographic drawings, which are developed following standard drawing practices. The process of adding size information to a drawing is known as dimensioning.

Geometric Dimensioning and Tolerancing (GD&T) is a way of defining and communicating engineering tolerances. The geometry, tolerances and other information such as surface finish, concentricity or parallelism are expressed using symbols and text. In turn, properly created engineering drawings communicate the degree of accuracy and precision needed on each controlled feature of the part.

Tolerancing specifications define the allowable variation for the form and possibly the size of individual features, and the allowable variation in orientation and location between features. Two examples are linear dimensions and feature control frames using a datum reference (both shown above). Standards define GD&T rules and drawing conventions include:

- o American Society of Mechanical Engineers (ASME) Y14.5-2009
- o International Organization for Standardization (ISO) ISO 2768:1989 General tolerances

- -Part 1: Tolerances for linear and angular dimensions without individual tolerance indications
- -Part 2: Geometrical tolerances for features without individual tolerance indications
- International Organization for Standardization (ISO) ISO 286:2010 Geometrical Product Specification

However, it should be noted that ASME Y14.5 standard provides a fairly complete set of standards for geometric dimensioning and tolerancing in one document. Where the ISO standards address a single topic at a time under different standards.

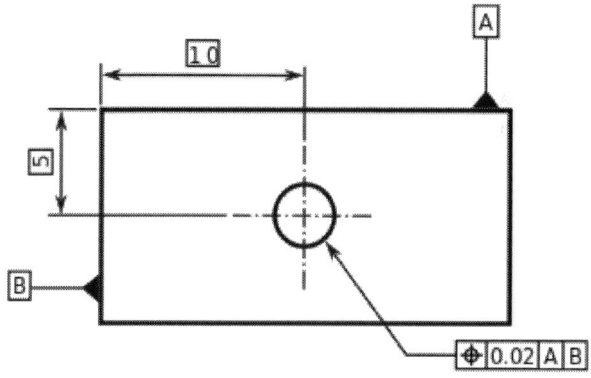

Figure: An example of geometric dimensioning and tolerancing (GD&T) of a hole. The cross signifies a position tolerance; the tolerance is 0.02 mm; the position is in reference to the datum planes A and B. The purpose of geometric dimensioning and tolerancing is to describe the engineering intent of (1) parts and/or (2) assemblies. To achieve this, some fundamental rules are applied:

- All dimensions must have a tolerance. Each feature on every manufactured part is subject to variation, therefore, the limits of allowable variation must be specified. Plus(+) and minus(-) tolerances may be applied directly to dimensions or alternatively by general note on the drawing or by a feature control frame.
- A dimension defines the nominal geometry and allowable variation.
- Engineering drawings define the requirements of finished (complete) parts. Additional dimensions would be helpful, but are not required, they may be marked as reference.
- The geometry should be described without explicitly defining the method of manufacture. Avoid describing manufacturing methods.
- All dimensioning and tolerancing should be assessed for readability and should be clearly visible.
- When geometry is normally controlled by gage sizes or by code (e.g. stock materials), the dimension(s) shall be included with the gage or code number in parentheses following or below the dimension.
- Angles of 90° are assumed when lines (including centre lines) are shown at right angles, but no angular dimension is explicitly shown.
- Dimensions and tolerances are valid at 20 °C / 101.3 kPa unless otherwise stated.
- Unless explicitly stated, all dimensions and tolerances are only valid when the item is in a free state.
- Dimensions and tolerances apply to the length, width, and depth of a feature including form variation.

Note: The rules above are not the exact rules stated in the ASME Y14.5-2009 standard.

Dimension: a numerical value that defines the size or geometric characteristic of a feature.

Basic dimension: a numerical value defining the theoretically exact size of a feature.

Reference dimension is the numerical value enclosed in parentheses provided for information only and is not used in the fabrication of the part.

Leader line: a thin solid line used to indicate the feature with which a dimension, note, or symbol is associated.

Datum: a theoretically exact point used as a reference for tabular dimensioning.

Tolerance: the amount a particular dimension is allowed to vary.

Dimension line: a thin solid line which shows the extent and direction of a dimension. Arrows are placed at the ends of dimension lines to show the limits of the dimension.

Extension line: a thin solid line perpendicular to a dimension line indicating which feature is associated with the dimension.

☐

CHAPTER 8
ISO 13485

Introduction

ISO 13485 is the quality management standard of choice for manufactures of medical devices. Revised in 2016, ISO 13485:2016 "specifies requirements for a quality management system where an organisation needs to demonstrate its ability to provide medical devices and related services that consistently meet customer and applicable regulatory requirements."1 The scope of the standard can apply to any organisation or company involved throughout the life-cycle of a product, including design and/or development, production, storage and distribution, installation, or servicing of a medical device and design and development or provision of technical or professional services.

The recent revision is designed to address recent developments in quality management and other updated regulations that relate to the industry. Improvements in the new version of the standard include broadening its applicability to include all organisations involved in the life cycle of the product, from the concept stage to end of life along with greater alignment with regulatory requirements and post-market surveillance including complaint handling.

ISO 13485:2016 is also used by suppliers or external vendors that provide QMS related management system- services. Requirements of ISO 13485:2016 are applicable to organisations regardless of their size and regardless of their type except where explicitly stated. Wherever requirements are specified as applying to medical devices, the requirements apply equally to associated services as supplied by the organisation. If any requirement in Clauses 6, 7 or 8 of ISO 13485:2016 is not applicable due to the activities undertaken by the organisation or the nature of the medical device for which the quality management system is applied, the organisation does not need to include such a requirement in its quality management system. For any clause that is determined to be not applicable, the organisation records the justification as part of their certification and quality management system.

The Process Approach

ISO 13485 is based on a process approach to quality management. A process is any activity that receives inputs and converts them to outputs. For an organisation to function effectively, it has to identify and manage numerous linked processes. Furthermore, many processes impact other processes or downstream processes. The application of a system of processes within an organisation, together with the identification and interactions of these processes, and their management, can be referred to as the "process approach".

Directives Versus Standards

When it comes to regulated industries such as medical devices, it is first important to be familiar with some common terms and definitions and what they really mean. This chapter examines some key terms that are applied widely and relate to regulated industries. They include:
- Directives
- Standards

- Notified Body
- Competent Authority

Directives

Directives are legal requirements which must be met by manufacturers or other bodies within the industry. Directives are based on legislation and are issued at governmental level. It is important to note that standards such as ISO 13485 help companies meet the requirements set up in directives. (See harmonised standards below)

Standards

Standards are not always mandatory. However, they help manufacturers be compliant with directives/legislation.
They also represent the current and best practice in the field of study/industry. Harmonised standards are European standards prepared under a mandate from the European Commission, referenced in the official journal, and drafted so that compliance with their requirements relates to one or more essential requirements of the directive. These standards have special status because, when a manufacturer can show that their products meet the requirements of the standard, there is a presumption that the product conforms to the essential requirements of the directive that is covered by the standard.

What is a Competent Authority?

When it comes to medical devices, a competent authority is the legally delegated authority mandated to monitor compliance to directives and legal requirements within the industry. The competent authority has the power to grant and revoke licenses.

Example of Competent Authorities:

- FDA (Food and Drug Administration) CFR Code of Federal Regulations – U.S.
- MHRA (Medicines and Healthcare Regulatory Agency - UK
- HPRA (Health Products Regulatory Agency) - Ireland
- JPAL (Japanese Regulations for Medical Devices) – Japan

What Is a Notified Body?

A notified body is a certification organisation which the national authority (the competent authority) of a member state designates to carry out one or more of the conformity assessment procedures described in the annexes of the medical devices directives. The Medicines and Healthcare Products Regulatory Agency is the UK competent authority under the three directives.

Organisations and Institutions

Many of the common acronyms that are referenced in literature relate to various standard setting organisations and industry representatives. Some of the more common bodies are listed below:

ISO: Internal Organisation for Standardisation
IMDR (F): International Medical Device Regulators Forum
ASTM: American Society for Testing and Materials
GHTF: Global Harmonisation Task Force

Basic Definitions (Source: Annex IX of Directive 93/42/EEC)

Intended Purpose: Intended purpose means the use for which the device is intended according to the data supplied by the manufacturer on the labelling, in the instructions and/or in promotional materials. (Chapter I section 1 of Annex IX of Directive 93/42/EEC)

Transient: Normally intended for continuous use for less than 60 minutes.

Short Term: Normally intended for continuous use for not more than 30 days.

Long Term : Normally intended for continuous use for more than 30 days.

Invasive Devices: A device which, in whole or in part, penetrates inside the body, either through a body orifice or through the surface of the body.

Body Orifice: Any natural opening in the body, as well as the external surface of the eyeball, or any permanent artificial opening, such as a stoma.

Surgically Invasive Device: An invasive device which penetrates inside the body through the surface of the body, with the aid of or in the context of a surgical operation.

Implantable Device: Any device which is intended:

- to be totally introduced into the human body or,
- to replace an epithelial surface or the surface of the eye, by surgical intervention which is intended to remain in place after the procedure. Any device intended to be partially introduced into the human body through surgical intervention and intended to remain in place after the procedure for at least 30 days is also considered an implantable device.

Medical Device: means any instrument, apparatus, appliance, material or other article, whether used alone or in combination, together with any accessories or software for its proper functioning, intended by the manufacturer to be used for human beings in the:

- diagnosis, prevention, monitoring, treatment or alleviation of disease or injury.
- investigation, replacement or modification of the anatomy or of a physiological process.
- control of conception which does not achieve its principal intended action by pharmacological, chemical, immunological or metabolic means.

A medical device may be assisted in its function by the following means:

Active Medical Device: any medical device relying for its functioning on a source of electrical energy or any source of power other than that directly generated by the human body or gravity.

Active Implantable Medical Device: any active medical device which is intended to be totally or partially introduced, surgically or medically, into the human body or by medical intervention into a natural orifice, and which is intended to remain after the procedure.

Custom-Made Device: means any active implantable medical device specifically made in accordance with a medical specialist's written prescription which gives, under his responsibility, specific design characteristics and is intended to be used only for an individual named patient.

Device Intended for Clinical Investigation: any active implantable medical device intended for use by a specialist doctor when conducting investigations in an adequate human clinical environment.

Intended Purpose: means the use for which the medical device is intended and for which it is suited according to the data supplied by the manufacturer in the instructions.

Putting into Service: means making available to the medical profession for implantation.

Where an active implantable medical device is intended to administer a substance defined as a medicinal product within the meaning of Council Directive 65/65/EEC of 26 January 1965 on the approximation of provisions laid down by law, regulation or administrative action relating to proprietary medicinal products (6), as last amended by Directive 87/21/EEC (7), that substance shall be subject to the system of marketing authorisation provided for in that directive.

Where an active implantable medical device incorporates, as an integral part, a substance which, if used separately, may be considered to be a medicinal product within the meaning of Article 1 of Directive 65/65/EEC, that device must be evaluated and authorised in accordance with the provisions of this directive.

ISO 13485 & Regulations

In chapter 2 the special status of harmonised standards was described which allows companies meet the essential requirements of Directives. In Europe, EN ISO 13485:2013 helps companies meet the requirements of: Directive 93/42/EEC on medical devices. This harmonised standard gives companies the "presumption of conformity" to complying with directives.

EN ISO 13485 was published in February 2013 and harmonised in August 2013 to cover the three directives:
90/385/ECC– The Active Implantable Medical Devices Directive (AIMDI)

> 93/42/ECC – The Medical Devices Directive (MDD)
> 98/79/EEC – In Vitro Diagnostic MDD (IVDMDD)

In the United States, medical device manufacturers need to meet the requirements of 21 CFR Part 820 of FDA regulations. While ISO 13485 is not an actual requirement, many companies will seek certification to the standard to support the exporting of products. In Australia, it is a regulatory requirement for manufacturers of medical devices to meet the requirements of ISO 13485.

In Canada, certification to ISO13485 is part of the regulatory requirements. The content of ISO 13485 is interpretive (not prescriptive) which gives a degree of scope in how the requirements are applied and met within a company. ISO 13485 provides both a sound and widely recognised basis in meeting regulatory compliance for medical devices. Based off ISO 19001 however, ISO 13485 is a standalone standard for medical devices.

ISO 9001 has requirements and themes relating to continual improvement and customer satisfaction. These have been modified for ISO 13485.

Main differences between ISO 9001 & ISO 13485:

- Customer satisfaction is changed to customer feedback
- Extra requirements regarding procedures for ISO 13485
- Extra requirements for records ISO 13485 (e.g. retention)
- Continual improvement is restricted to continual improvement of the quality management system

ISO 13485 has extra requirements required for regulatory bodies such as post production review and management of advisory events.

ISO 13485 and ISO/TR 14969

ISO/TR 14969 is a technical report that is used for guidance on the application and implantation of ISO 13485. It is recommended for those responsible for the role out of ISO 13485 within their organisation. The content of ISO/TR 14969 is based on several established organisations such as the GHTF, ISO and input from regulatory bodies.
Standard Overview

ISO 13485 has 8 Clauses or Sections which make up the structure of the standard.

Section 0 Normative References, Definitions and Terms
Section 1 Requirements of the Quality Management System (QMS)
Section 2 Normative References
Section 3 Terms and Definitions
Section 4 Requirements of the Quality Management System (QMS)
Section 5 Management Responsibility
Section 6 Resource Management
Section 7 Product Realisation
Section 8 Measurement, Analysis and Improvement

CLAUSE 1: SCOPE

This section refers to the scope and application of the standard.

The organisation must be able to show its ability to provide medical devices to meet customer requirements and regulatory requirements. A key aim of the standard is to allow harmonisation to regulatory requirements.

The scope of the QMS must relate to medical devices for a company to be able to use ISO 13485.

Some examples of what's in scope of the standard include (1) the manufacture of hip implants, (2) the design and manufacturing of in-vitro blood testing devices, (3) contact analytical testing (4) consultancy services. The terms "where appropriate" and "if appropriate" are used throughout the standard, therefore, it should be met by the organisation unless a justification is documented.

CLAUSE 2: NORMATIVE REFERENCES

This clause states that when working with ISO 13485, refer to ISO 9000:2000 for fundamentals and vocabulary.

CLAUSE 3: TERMS AND DEFINITIONS

This clause provides terms and definitions. It is very useful in the early days of establishing and implementing ISO 13485 to ensure that terms and definitions are clearly understood.

CLAUSE 4: QUALITY MANAGEMENT SYSTEM
Clause 4 details the general requirements that relate to the quality management system, the documentation requirements and record requirements.

Clause 4 includes:
4.1 General requirements clause
4.2 Documentation requirements clause

CLAUSE 4.1 GENERA REQUIREMENTS

The organisation must implement a Quality Management System, or QMS in order to provide the framework and structure to achieve ISO 13485 roll-out and implementation. However, the role of the QMS does not stop there. After initial roll-out, the requirements of the standard must be maintained and determined to be effective on an on-going basis. The following processes should be documented:

- List of all processes
- Process interactions
- Monitoring of processes
- Resources to facilitate rollout of processes
- Measure and monitor effectiveness
- System of identifying improvements

CLAUSE 4.2: DOCUMENTATION REQUIREMENTS

When it comes to the regulated industries such as the medical device industry, every process and procedure must be documented. Documentation ensures that everyone is working in the same manner with the same procedures. However, documentation is more than just writing down procedures and processes. It is also concerned with how documents are controlled, how they are updated and how they are stored.
Electronic Document Management Systems

Electronic document management systems aka EDMS are now the norm and gold standard for most medium to large organisations. Many companies that provide medical device manufacturers with an EDMS can customise the system to match the business processes particular to an organisation. With configurable or customisable software, validation and proper verification is important to ensure the system operates as intended. There are also regulatory requirements that stipulate the expectations and requirements of such systems. For example, the application of electronic signatures and the presence of audit trials. FDA 21 CRF Part 11 details the requirements with regards to electronic records and electronic signatures. For medicinal products in Europe, GMP V4 Annex 11 specifies similar requirements.

Changes and Updates to Documents

Revision control is a key element of the Quality Management Systems in place in regulated industries. As the need for changes in the document arises, the controlled document can be amended/updated. With each update the version number revises also. Some companies will use alphabetic revision control and to a lesser extent numeric revision control (Version A, Version B or Version 01, Version 02).

Controlled documents should always have a version number or revision number electronically on each page of the document. This is similar to books which always list what edition they are. e.g. first edition or second edition.

Records

Records are generated through the application of processes and procedures. These records can be related in quality inspection and manufacturing. The integrity and quality of records relating to the manufacture of medical devices is important, as it plays a part in safe-guarding the patient or user. Records may also help in the investigation of any quality issues, complaints or adverse events that may arise.

Principles of Good Documentation Practices or GDP, should be applied to records. In particular, handwritten entries should always be accompanied by a signature and date. This is important as traceability must be maintained in the event of an issue or complaint.

CLAUSE 5: MANAGEMENT RESPONSIBILITY

Clause 5 includes:

5.1 Management Commitment
5.2 Customer Focus
5.3 Quality Policy
5.4 Planning
5.5 Responsibility, Authority and Communication Management Review

5.1 Management Commitment

It is essential that top management have an authentic and tangible commitment to meeting regulations and the expectations of customers. Quality should be at the forefront of all of activities. Management should encourage discourse and communication on all matters relating to internal processes, quality and the QMS as a whole.

5.2 Customer Focus

Customer –patient/user/doctor/family member
Customer feedback is a requirement of ISO 13485 and as such the manufacturer must engage with the customer. In instances where a defective product is received, the manufacturer must have a complaints process to facilitate proper feedback, communication and investigation.

5.3 Quality Policy

Simple statement /1 pager or more
Often quality policies will be displayed in reception areas etc. Copies should be signed and revision controlled.
Quality policy must have a commitment to maintain the effectiveness of the QMS.

5.4 Planning

Top management must plan quality objectives and ensure they are implemented and effective.
Some examples of quality objectives include:
- reduce rework by 10%
- reduce scrap by 5%
- have customer complaints reduced by 2% per year

5.5 Responsibility, Authority and Communication

Roles and responsibilities are defined.
Job descriptions are in place.
Organisational charts are in place and accurate.

5.6 Management Review

The purpose of management review is to ensure the effectiveness of the QMS.

Inputs to management review include:

(a) Audit results
(b) Customer feedback
(c) Process performance and conformity
(d) Corrective and preventative actions
(e) Deviations
(f) Regulatory changes and revisions

CLAUSE 6 : RESOURCE MANAGEMENT

Clause 6 of ISO 13485 is concerned with human resources, infrastructure and work environment.

Clause 6 includes:
6.2 Human resources
6.3 Infrastructure
6.4 Work Environment

People are the key part of any QMS. Therefore, they should have the appropriate level of education, skill and experience. A culture of quality must be lived by everyone.

People must be suitably trained. Training must be documented and consistent throughout an organisation. Training must be seen to be effective. Proper records of education and training must be kept.

Human intelligence, human creativity and human labour are all key inputs to any factory or company manufacturing medical devices. Therefore, an organisation must be properly resourced in order to function correctly, meet the regulatory requirements and customer expectations.

6.2 Human resources

"Change the people or change the people"

With any organisation, it is only as good as the people it has in its make-up. Therefore, the people, operators, engineers, managers etc. all contribute to the quality management system. Clause 6.2.2 also specifies requirements with regards to competence, awareness and training. The person should be matched to the job in terms of their qualifications, experience and training. Typically, job descriptions are used to drive and capture these requirements. Nowadays, most multinational companies will ask for evidence of qualifications, training and experience. These documents are then held on file in the event of an audit. This is recommended practice for medical device companies. While the standard does not specifically call out the need to hold records of degrees and qualifications on file, the company or organisation needs to demonstrate the suitability of the person to their respective roles, and filing the qualification provides the easiest method.

6.3 Infrastructure

Infrastructure has the ability to impact the quality of products and services. Therefore, it must be fit for purpose. It is especially important if the organisation is involved with the manufacture of medical devices. The following element need to be considered with regards to infrastructure:
- o Location of equipment and the operating environment
- o Equipment installation and validation
- o Utilities required for the operation of equipment and systems
- o Layout of the factory – flow or raw materials, in-process materials and finished products
- o Environmental systems such as HVAC and fire suppression systems

6.4 Work Environment

The work environment is also closely related to infrastructure within a given organisation and they can both affect or impact upon the quality of products manufactured. Risk to product quality and patients is minimised by understanding the work environment and how it can impact the product. When the interactions and risks are understood, work can then be done to eliminate risks or at least control or monitor them. Environmental conditions that can impact upon product quality include:

- o Humidity
- o Temperature
- o Air quality
- o Room pressure differentials (negative / positive)
- o Air flow/velocity

CLAUSE 7: PRODUCT REALISATION

Clause 7 includes:
7.1 Planning of product realisation
7.2 Customer-related processes
7.3 Design and development
7.4 Purchasing
7.5 Production and service provision
7.6 Control of measuring devices
7.1 Planning of Product Realisation

Product realisation can be defined as a collection of processes and body of work that delivers a product or service to the customer. Remember, when it comes to medical devices, customers can be patients or users such as doctors and nurses. It should be noted that organisations can opt to exclude specific requirements, in cases where product realisation is not applicable. However, any such exclusion should be based on sound rationale with the case clearly documented. An example of this may be where design and development is not conducted by the manufacturer e.g. contract manufacturers.

7.1 Planning of Product Realisation

Planning is an often underestimated but remains a key element of product realisation. If adequate time and resources are given to planning, it makes all other processes run smoother, and therefore should help to produce improved products and services.

7.2 Customer-Related Processes

There are 3 elements that feed into customer-related processes. They include the following:

Determining the requirements related to the product Clause 7.2.1
Review of requirements relating to the product-Clause 7.2.2
Customer communication-clause 7.2.3

Customer requirements are typically captured in a User Requirements Specification. A requirements specification (URS) documents all of the desired attributes of a product or service. They can be made up by a combination of CQAs, regulatory requirements and design requirements. A URS can then form the basis for review of the product or service requirements.

With regard to customer communication, it is important to remind ourselves that we are concerned with ISO 13485 which as we very well know by now is the standard for medical devices. Therefore, having the right information available to the customer, patient or end user is important. When additional information needs to be transmitted or updates to information need to be communicated, an advisory note can be issued. Another important aspect of customer communication is customer feedback. This communication can be made up of positive feedback from the customer or users, or when there is a query with regard to a product or service. Therefore, processes or systems must be in place to make communication between customer and company both effective and timely.

7.3 Design and Development

Design and Development Verification and Validation ensure that the product is designed, developed and subsequently manufactured meeting all the customer requirements, regulatory requirements and business requirements. These requirements are classed as inputs to the design and development, and verification and validation ensure the inputs have been adequately taken into account.

The design and development testing sometimes replicate the commercial applications of the medical device, hence providing a realistic challenge in order to have confidence in the medical device.

Design Control

Design control is a necessary practice that ensures good engineering principles are maintained throughout the design phase of a product. It also refers to the continual design and development of the product through its very lifecycle. The design and development files and history must be controlled and maintained, with any changes properly assessed, tested and documented.

7.4 Purchasing

Bearing in mind that a quality management system considers all aspects of an organisation's functioning, purchasing and procurement of materials necessitates putting robust controls in place. Simply put, a purchasing process must exist.

7.5 Production and Service Provision

This requirement of ISO 13485 is an extensive section with a great deal of importance associated with it. As we are dealing with the manufacture of medical devices (or other activity associated with medical devices) there are specific requirements for sterile products. If a product is sterile, its use or application is likely to be associated with greater risks to the patient. Therefore, extra safeguards must be in place for sterile medical devices. Key sections of Clause 7.5 include: (1) control of production and service provision – both general and specific requirements, (2) specific requirements for sterile medical devices, (3) validation of equipment and processes for production and service provision, (4) traceability and identification, (5) preservation of product controls with regard to monitoring and measuring medical devices.

7.6 Control of Measuring Devices

This clause requires an organisation to identify what monitoring and measuring is required and to ensure the product or service meets the customer requirements. A calibration procedure must also be maintained to ensure the equipment is accurate and reliable. Calibration must ensure that:
Equipment used to verify product quality is calibrated to a periodic schedule.
- o The calibration is performed to an international standard.
- o The calibration status of the equipment is recorded and visible.
- o The equipment must be located within a suitable area in order to maintain accurate and reliable results.

If an organisation uses any computer software to monitor or measure outputs, the software must be verified before use via the appropriate validation and qualification activities.

CLAUSE 8: MEASUREMENT ANALYSIS

Clause 8 includes:
8.1 General requirements
8.2 Monitoring and measurement
8.3 Control of nonconforming products
8.4 Analysis of data
8.5 Improvement

8.1 General Requirements

Measurement, analysis and improvement are the key themes of clause 8. As with all medical devices, inspection and testing both during manufacturing and post manufacturing is necessary to ensure products and services function as intended and without defects. With any type of measurement or inspection analysis, the method used to complete the testing is critical. The method must be fit for purpose, and the equipment must be suitable. This "method validation" typically is done during the design and development phase.

8.2 Monitoring and Measurement

Monitoring and measurement are dependent on the information or feedback provided from various sources. The most important feedback is the post-production feedback that is gathered from customers or the end user. Again, this occurs over the whole lifetime of the product or service in question. There are a number of methods that can be used to obtain feedback. Some examples include:

-Customer surveys
-Customer complaints
-Review of regulatory databases such as MAUDE.
-Repair and servicing information

8.3 Control of Nonconforming Product

Non-conforming product presents a risk to patients or users of medical devices. When a situation arises where non-conforming product is manufactured or detected through inspection processes, the product must be controlled and segregated to prevent unintended use or distribution.

Some examples resulting in non-conformance are:
- o When a manufacturing process drifts outside its validation window or operating parameters.
- o A certificate of analysis for a raw material is not provided by the supplier or the results are out of specification.
- o In-process testing was not completed at the defined intervals.
- o Training of personnel completing tests is not current or is inadequate.

8.4 Analysis of Data

In any engineering activity, data and the quality of the data is a key factor in making the right decisions. Provided the data collected is relevant and accurate, analysis of data can provide important insights into process performance, quality control and product functionality. Data should be collated in a consistent way and controlled by a procedure. When it comes to medical device manufacturing, the sources and types of data are multiple. Data can be generated from in-process testing and data can be generated from end of line testing aka finished product testing.

8.5 Improvement

ISO 13485 fosters a culture of continual improvement. As we have seen, each activity can be described as a process. For example, a manufacturing process, a procurement process, a complaints process. The set of processes that make up the quality management system need to be continually reviewed to ensure they are suitable and effective for the task at hand. Typical tools used to keep improvement in mind include:

- Review of the quality policy and quality objectives
- Frequency and category of corrective and preventative actions (CAPA's)
- Customer complaints
- Management review input

CE Marking

In Europe a QMS is required for CE marking of a medical device that is placed on the market in the EU.
ISO 13485:2003 is a harmonised standard that can be used by companies to show conformity of their QMS to requirements of directives. EN ISO 13485:2012 was harmonised in August 2012. This allows compliant companies receive an EC Declaration of Conformity.

Summary of the CE Requirements

Manufacturers of class I devices or their authorised representatives must:
- review the classification rules to confirm that their products fall within class I (Annex IX of the Directive)
- check that their products meet the essential requirements (Annex I of the Directive)

- notify the competent authority, in advance, of any proposals to carry out a clinical investigation to demonstrate safety and performance of a device as required by the regulations
- obtain notified body approval for sterility or metrology aspects of their devices and where applicable prepare relevant technical documentation
- Draw up the 'EC Declaration of Conformity' (below) before applying the CE marking to their devices
- Register with the competent authority
- Implement and maintain corrective action and vigilance procedures including a systematic procedure to review experience gained in the post-production phase
- Make available relevant documentation on request for inspection by the competent authority.

In Europe, all medical devices must bear the CE marking of conformity (see Annex XII) of the directive) when they are placed on the market and/or put into service. The CE marking must appear in a visible, legible and indelible form on the device or its sterile pack, where practicable and appropriate, and where applicable on any instructions for use and sales packaging. For 'sterile' and 'measuring' devices, the CE marking must be accompanied by the identification number of the notified body that has acted under the relevant conformity assessment procedure.

EC Declaration of Conformity

In order to affix the CE marking, the manufacturer or their authorised representative must follow the EC declaration of conformity procedure referred to in Annex VII of the directive. This procedure must be completed prior to placing the device on the market. The 'EC declaration of conformity' is the procedure whereby the manufacturer or their authorised representative prepares the required technical documentation, puts into place corrective action and vigilance procedures and declares that the products meet the requirements set out in the directive.

Technical Documentation

The technical documentation should be prepared following review of the essential requirements and must cover all of the following aspects:

Description: A general description of the product, including any variants (for example names, model numbers and sizes).
Raw Materials and Component Documentation: Specifications including, as applicable, details of raw materials, drawings of components and/or master patterns and any quality control procedures.

Intermediate Product and Sub-Assembly Documentation: Specifications including appropriate drawings and/or master patterns, circuits, and formulation specification; relevant manufacturing methods and any quality control procedures.
Packaging and Labelling Documentation: Packaging specifications and copies of all labels and any instructions for use.
Design Verification: The results of qualification tests and design calculations relevant to the intended use of the product, including connections to other devices in order for it to operate as intended.

Risk Analysis: The results of risk analysis to review whether any risks associated with the use of the product are compatible with a high level of protection of health and safety and are acceptable when weighed against the benefits to the patient or user. If biocompatibility is relevant – for example for skin contact and invasive devices – a compilation and review of existing data or test reports based on the relevant standards is required.

Compliance with the Essential Requirements and Harmonised Standards: A list of relevant harmonised standards (for example sterilisation, labelling and information, biocompatibility, electrical safety, risk analysis, product group standards) which have been applied in full or in part of the products. If relevant harmonised standards have not been applied in full, then additional data will be required, detailing the solutions adopted to meet the relevant essential requirements of the directive. The manufacturer may choose to prove conformity with the essential requirements of the directive through the use of their own standards and/or other relevant published standards (ISO, EN, BS). However, the use of such standards does not give similar, immediate presumption of conformity to the essential requirements of the directive. Therefore, using a harmonised standard provides greater protection to the manufacturer.

Device Classification

The manufacturer, in preparing for CE marking, should first determine if their product falls within the scope of the directive or national regulation, either as a medical device or as an accessory to a medical device, as defined in Article 1 of directive 93/42/EEC and Article 2 of the regulation. In order to be classified as a medical device, the product should have a medical purpose and its primary mode of action will typically be physical.

General medical devices and related accessories must be classified into one of four classes, which are based on the perceived risk of the device to the patient or user. The classification of a device determines the conformity assessment options that are applicable to the device, with higher risk devices undergoing higher levels of assessment.

Class	Risk level
I	Low Risk
IIa	Medium Risk
IIb	Higher Risk
III	Highest Risk

Classification Rules

There are eighteen rules outlined in Annex IX of the directive and related regulation that lay down the basic principles of classification. In MEDDEV 2.4/1 Rev. 8, these rules are further explained and descriptive examples are provided.

The eighteen rules are subdivided into four groups as follows:

Rules   Device Type
Rules 1 – 4   Non-invasive Devices
Rules 5 – 8   Invasive Devices
Rules 9 – 12   Active Devices
Rules 13-18   Special Rule e.g. devices containing tissue of animal origin, drug-device combinations
Table: Rules Corresponding to Device Type.

Annex IX and related guidance documents outline a number of key characteristics, listed below, that must be considered to correctly classify a device using the eighteen classification rules:

General Principles of Device Classification

Medical devices are defined as articles which are intended to be used for a medical purpose. It is the intended purpose that determines the class of device and not the particular technical characteristics of the device. The intended purpose of the device should be substantiated (if required) and be representative of the technical characteristics of the device.
It is the intended and not the accidental use of the device that determines its class.
It is the intended purpose assigned by the manufacturer to the device that determines the class of device and not the class assigned to other similar products.
Accessories are classified separately from their parent device.
The mode of action of a medical device should be clear and evidenced with appropriate data to confirm this mode of action.
If the device can be classified according to several rules then the highest possible class applies.
Multipurpose equipment which may be used in combination with medical devices are not themselves classed as medical devices unless the manufacturer places them on the market with the specific intended purpose as a medical device.

If the device is not intended to be used solely or principally in a specific part of the body, it must be considered and classified on the basis of the most critical specified use.

Summary Of Rules

(Source: Guidelines Relating To The Application Of The Council Directive 93/42/EEC On Medical Devices, MEDDEC 2.4/Rev.9 June 2010)

Rule 1

Rule 1: All non-invasive devices are in Class I, unless one of the other 17 rules apply. This is a fallback rule applying to all devices that are not covered by a more specific rule.

This is a rule that applies in general to devices that come into contact only with intact skin or that do not touch the patient.

Some non-invasive devices are indirectly in contact with the body and can influence internal physiological processes by storing, channeling or treating blood, other body liquids or liquids which are returned or infused into the body or by generating energy that is delivered to the body. These must be excluded from the application of this rule and be handled by another rule because of the hazards inherent in such indirect influence on the body.

Rule 2

Rule 2: All non-invasive devices are in Class I, unless one of the other 17 rules apply.

These types of devices must be considered separately from the non-contact devices of Rule 1 because they may be indirectly invasive. They channel or store substances that will eventually be administered to the body. Typically these devices are used in transfusion, infusion, extracorporeal circulation and delivery of anaesthetic gases and oxygen.

In some cases devices covered under this rule are very simple gravity activated delivery devices.

Rule 2: All non-invasive devices intended for channelling or storing blood, body liquids or tissues, liquids or gases for the purpose of eventual infusion, administration or introduction into the body are in Class IIa:
- if they may be connected to an active medical device in Class IIa or a higher class,

-if they are intended for use for storing or channelling blood or other body liquids or for storing organs, parts of organs or body tissues.

- in all other cases they are in Class I.

Rule 3

Rule 3: Non-invasive devices that modify biological or chemical composition of blood, body liquids or other liquids intended for infusion into the body.

These types of devices must be considered separately from the non-contact devices of Rule 1 because they are indirectly invasive. They modify substances that will eventually be infused into the body. This rule covers mostly the more sophisticated elements of extracorporeal circulation sets, dialysis systems and autotransfusion systems as well as devices for extracorporeal treatment of body fluids which may or may not be immediately reintroduced into the body, including, where the patient is not in a closed loop with the device.

Rule 3: All non-invasive devices intended for modifying the biological or chemical composition of blood, other body liquids or other liquids intended for infusion into the body are in Class IIb,
unless the treatment consists of filtration, centrifugation or exchange of gas or heat, in which case they are in Class IIa.

These devices (Rule 3) are normally used in conjunction with an active medical device covered under Rule 9 or Rule 11.
Filtration and centrifugation should be understood in the context of this rule as exclusively mechanical methods.

Rule 4

Rule 4: Non-invasive devices which come into contact with injured skin.
This rule is intended to primarily cover wound dressings independently of the depth of the wound. The traditional types of products, such as those used as a mechanical barrier, are well understood and do not result in any great hazard. There have also been rapid technological developments in this area, with the emergence of new types of wound dressings for which non-traditional claims are made, e.g. management of the micro-environment of a wound to enhance its natural healing mechanism.

More ambitious claims relate to the mechanism of healing by secondary intent, such as influencing the underlying mechanisms of granulation or epithelial formation or preventing contraction of the wound. Some devices used on breached dermis may even have a life-sustaining or lifesaving purpose, e.g. when there is full thickness destruction of the skin over a large area and/or systemic effect.
Dressings containing medicinal products which act ancillary to the dressing fall within Class III under Rule 13.

Rule 4: All non-invasive devices which come into contact with injured skin:

- are in Class I if they are intended to be used as a mechanical barrier, for compression or for absorption of exudates,

- are in Class IIb if they are intended to be used principally with wounds which have breached the dermis and can only heal by secondary intent.

Products covered under this rule are extremely claim sensitive, e.g. a polymeric film dressing would be in Class IIa if the intended use is to manage the micro-environment of the wound or in Class I if its intended use is limited to retaining an invasive cannula at the wound site. Consequently it is impossible to say a priori that a particular type of dressing is in a given class without knowing its intended use as defined by the manufacturer. However, a claim that the device is interactive or active with respect to the wound healing process usually implies that the device is in Class IIb.

Most dressings that are intended for a use that is in Class IIa or IIb, also perform functions that are in Class I, e.g. that of a mechanical barrier. Such devices are nevertheless classed according to the intended use in the higher class.

For such devices incorporating a medicinal product or a human blood derivative see Rule 13 or animal tissues or derivatives rendered non-viable see Rule 17.

Rule 5

Rule 5: Devices invasive with respect to body orifices.

Invasiveness with respect to the body orifices (ear, mouth, nose, eye, anus, urethra and vagina) must be considered separately from invasiveness that penetrates through a cut in the body surfaces (surgical invasiveness). For short term use, a further distinction must be made between invasiveness with respect to the less vulnerable anterior parts of the ear, mouth and nose and the other anatomical sites that can be accessed through natural body orifices.

Surgically created stoma, which for example allows the evacuation of urine or faeces, should also be considered as a body orifice.

Devices covered by this rule tend to be diagnostic and therapeutic instruments used in particular specialities (ENT, ophthalmology, dentistry, proctology, urology and gynaecology).

Rule 5: All invasive devices with respect to body orifices, other than surgically invasive devices and which are not intended for connection to an active medical device or which are intended for connection to an active medical device in Class I:
- are in Class I if they are intended for transient use,
- are in Class IIa if they are intended for short term use
except if they are used in the oral cavity as far as the pharynx, in an ear canal up to the ear drum or in a nasal cavity, in which case they are in Class I,
- are in Class IIb if they are intended for long term use,
except if they are used in the oral cavity as far as the pharynx, in an ear canal up to the ear drum or in a nasal cavity and are not liable to be absorbed by the mucous membrane, in which case they are in Class IIa.

All invasive devices with respect to body orifices, other than surgically invasive devices, intended for connection to an active medical device in Class IIa or a higher class, are in Class IIa.

Rule 6

Rule 6: Surgically invasive devices intended for transient use (< 60 minutes)

This rule primarily covers three major groups of devices: devices that are used to create a conduit through the skin (needles, cannulae, etc.), surgical instruments (scalpels, saws, etc.) and various types of catheters, suckers, etc.

This rule primarily covers three major groups of devices: devices that are used to create a conduit through the skin (needles, cannulae, etc.), surgical instruments (scalpels, saws, etc.) and various types of catheters, suckers, etc.

Rule 6: All surgically invasive devices intended for transient use are in Class IIa unless they are:
-intended specifically to control, diagnose, monitor or correct a defect of the heart or of the central circulatory system through direct contact with these parts of the body, in which case they are in Class III
-reusable surgical instruments, in which case they are in Class I
-intended specifically for use in direct contact with the central nervous system, in which case they are in Class III,
- intended to supply energy in the form of ionising radiation in which case they are in Class IIb,
- intended to have a biological effect or to be wholly or mainly absorbed in which case they are in Class IIb,
- intended to administer medicines by means of a delivery system, if this is done in a manner that is potentially hazardous taking account of the mode of application, in which case they are Class IIb.

Rule 7

Rule 7: Surgically invasive devices intended for short-term use (>60 minutes, <30 days).
These are mostly devices used in the context of surgery or post-operative care (e.g. clamps, drains), infusion devices (cannulae, needles) and catheters of various types.

Rule 7: All surgically invasive devices intended for short term use are in Class IIa unless they are intended:
- either specifically to control, diagnose, monitor or correct a defect of the heart or of the central circulatory system through direct contact with these parts of the body, in which case they are in Class III,

- or specifically for use in direct contact with the central nervous system, in which case they are in Class III,
- or to supply energy in the form of ionising radiation in which case they are in Class IIb,
- intended to have a biological effect or to be wholly or mainly absorbed in which case they are in Class III, - or to undergo chemical change in the body, except if the devices are placed in the teeth, or to administer medicines, in which case they are Class IIb.

Rule 8

Rule 8: Implantable devices and long-term surgically invasive devices (> 30 days). These are mostly implants in the orthopaedic, dental, ophthalmic and cardiovascular fields as well as soft tissue implants such as implants used in plastic surgery.

Rule 8: All implantable devices and long-term surgically invasive devices are in Class IIb unless they are intended:
- to be placed in the teeth, in which case they are in Class IIa,
- to be used in direct contact with the heart, the central circulatory system or the central nervous system, in which case they are Class III,
- to have a biological effect or to be wholly or mainly absorbed, in which case they are in Class III,
- or to undergo chemical change in the body, except if the devices are placed in the teeth, or to administer medicines, in which case they are in Class III.
- Directive 2003/12/EC introduced a derogation from this rule, reclassifying breast implants in Class III
Directive 2005/50/EC introduced a derogation from this rule, reclassifying hip, knee and shoulder joint replacements in Class III

Rule 9

Rule 9: Active therapeutic devices intended to administer or exchange energy.

Devices classified by this rule are mostly electrical equipment used in surgery such as lasers and surgical generators. In addition there are devices for specialised treatment such as radiation treatment. Another category consists of stimulation devices, although not all of them can be considered as delivering dangerous levels of energy considering the tissue involved.

Rule 9: All active therapeutic devices intended to administer or exchange energy are in Class IIa
unless their characteristics are such that they may administer or exchange energy to and from the human body in a potentially hazardous way, taking account of the nature, the density and the site of application of the energy, in which case they are in Class IIb. All active devices intended to control or monitor the performance of active therapeutic devices in Class IIb or intended to influence directly the performance of such devices are in Class IIb.

Rule 10

Rule 10: Active devices for diagnosis. This primarily covers a whole range of widely used equipment in various fields, e.g. ultrasound diagnosis, capture of physiological signals and therapeutic and diagnostic radiology.

Rule 10: Active devices intended for diagnosis are in Class IIa:

- if they are intended to supply energy which will be absorbed by the human body, except for devices used to illuminate the patient's body, in the visible spectrum,

- if they are intended to image in vivo distribution of radiopharmaceuticals,
- if they are intended to allow direct diagnosis or monitoring of vital physiological processes,

unless they are specifically intended for monitoring of vital physiological parameters, where the nature of variations is such that it could result in immediate danger to the patient, for instance variations in cardiac performance, respiration, activity of CNS in which case they are in Class IIb.

Active devices intended to emit ionising radiation and intended for diagnostic and therapeutic interventional radiology including devices which control or monitor such devices, or which directly influence their performance, are in Class IIb.

Rule 11

Rule 11: Active devices intended to administer and/or remove medicines, body liquids or other substances to or from the body. This rule is intended to primarily cover drug delivery systems and anaesthesia equipment.

Rule 11: All active devices intended to administer and/or remove medicines, body liquids or other substances to or from the body are in Class IIa, unless this is done in a manner:
- that is potentially hazardous, taking account of the nature of the substances involved, of the part of the body concerned and of the mode of application, in which case they are in Class IIb.

Rule 12

Rule 12: All other active devices. This is a fall-back rule to cover all active devices not covered by the previous rules.

Rule 12: All other active devices are in Class I

Special Rules 12-18

Rule 13: Devices incorporating, as an integral part, a medicinal product or a human blood derivative (See MEDDEV. 2.1/3 for further guidance).
Rule 14: Devices used for contraception or prevention of sexually transmitted diseases.
Rule 15: Specific disinfecting, cleaning and rinsing devices.
Rule 16: Devices to record X-ray diagnostic images.
Rule 17: Devices utilising animal tissues or derivatives.
Rule 18: Blood bags.

CHAPTER 9
LEAN BASICS

Introduction

Lean is a globally recognised set of principles and practices that help build and continually improve businesses and organisations across different industries and sectors. Lean can be applied to small projects and large projects that are delivered over the short, medium or long-term. The origins of lean date back to over 50 years ago to the Toyota production system. This was an in-house methodology developed within the Toyota manufacturing company of Japan. From the early 1980s lean principles began to become more widely known and acknowledged and very quickly its popularity grew to impact different sectors in different countries. It is without question a proven way to ensure businesses are more effective and customer focused while maximising value and quality.

Applying lean techniques can help companies reduce defects and deficiencies and most of all continually improve their systems and processes. It also works to eliminate waste of materials and wasting time and other resources.

What is lean?

If you are new to lean in a manufacturing environment, you may question what is the essential meaning of the lean philosophy? At its core is a continual drive to do more with less. If Lean is correctly implemented, a company or organisation will (1) use less human effort to perform their work, (2) use less material and (3) manufacture less defects. Not only will lean achieve these goals, it also works to maintain and sustain the results over time.

Key Points of Lean:
- o A focus on customer value
- o Getting the whole team involved
- o A philosophy of continuous improvement
- o Reducing variation
- o Eliminating waste
- o Taking the long-term view
- o Improving value
- o Providing exactly what's needed at the right time based on customer demand and requirements maintaining flow and the right movement at the right time

A Roadmap to Lean

(1) The true value of a product or service is must be based on the customer's perspective. This is an important point to grasp. Value is not what an individual engineer or department deems it to be. It must be established from the customer. Therefore, the first step should be to clearly specify value as the customer sees it.
(2) In identifying all the steps within a value stream, this will allow the team to determine if steps are value adding or non-value adding.
(3) Make only those actions which create value flow.
(4) Only make what is pulled by the customer just-in-time.

(5) Perfection is the goal. Strive towards perfection by continually removing successive layers of waste.

The History of Lean

The early beginnings of lean can be dated right back to the 1900s. What was referred to as Time & Motion studies, pioneered by an American Engineer Frederick Winslow Taylor. It was one of the first business efficiency techniques (even this term wasn't yet defined). The idea of time and motion studies is to breakdown large tasks into small steps in the sequence they occur and the exact time taken for each "motion" or step to be completed.

Henry Ford focused on reducing waste while developing the mass assembly manufacturing system for car production. This awareness of waste as a drain goes back to as early as 1915. Henry Ford himself, quoting from in My Life and Work (1922), provided a single-paragraph description that encompasses the entire concept of waste:

"I believe that the average farmer puts to a really useful purpose only about 5% of the energy he expends. Not only is everything done by hand, but seldom is a thought given to a logical arrangement. A farmer doing his chores will walk up and down a rickety ladder a dozen times. He will carry water for years instead of putting in a few lengths of pipe. His whole idea, when there is extra work to do, is to hire extra men. He thinks of putting money into improvements as an expense.... It is waste motion— waste effort— that makes farm prices high and profits low." Ford looked at the bigger picture and understood that poor arrangement of the workplace leads to waste and inefficiencies, such as the length of time to transport water manually.

Design for Manufacture (DFM) - a Ford Concept

This standardisation of parts was central to Ford's concept of mass production, and the manufacturing "tolerances", or upper and lower dimensional limits that ensured interchangeability of parts became widely applied across manufacturing. Decades later, the renowned Japanese quality guru, Genichi Taguchi, demonstrated that this "goal post" method of measuring was inadequate.

He showed that "loss" in capabilities did not begin only after exceeding these tolerances, but increased as described by the Taguchi Loss Function at any condition exceeding the nominal condition. This became an important part of W. Edwards Deming's quality movement of the 1980s, later helping to develop improved understanding of key areas of focus such as cycle time variation in improving manufacturing quality and efficiencies in aerospace and other industries.

While Ford is renowned for his production line it is often not recognised how much effort he put into removing the fitters' work to make the production line possible. Until Ford, a car's components always had to be fitted or reshaped by a skilled engineer at the point of use, so that they would connect properly. By enforcing very strict specification and quality criteria on component manufacture, he eliminated this work almost entirely, reducing manufacturing effort by between 60-90%. However, Ford's mass production system failed to incorporate the notion of "pull production" and thus often suffered from over-production.

Dr. Deming's Management System

The contribution of Dr. Deming to Lean and modern manufacturing was most significant during the 1950s. With a strong background in statistics, Deming studied at New York University's graduate school of business.

Deming became acquainted with Walter A. Shewhart of the Bell Telephone Company. Shewhart was seen as a pioneer in the fields of statistical control of processes and the tools of the control charts. His influence led to Deming becoming interested in using statistical methods in the fields of manufacturing, industrial production and management. Furthermore, Shewhart's idea of common and special causes of variation resulted in the formation of Deming's theory of management. During the summer of 1950, Deming lectured hundreds of engineers, managers, and academics in statistical process control (SPC) and the concepts of quality.

It is also believed that top management from key Japanese companies were also trained by Deming during this time. The implementation of Deming's systems led to improved quality and lower costs. This led to an increased demand for Japanese products across the globe.

Ford Motor Company was one of the first American corporations to seek help from Deming. Deming questioned the company's culture and the way its managers operated. To Ford's surprise, Deming did not talk about quality, but about management. He told Ford that management actions were responsible for 85 percent of all problems in developing better cars. By 1986, Ford had become the most profitable American auto company. In 1982, Deming's book Quality, Productivity, and Competitive Position was published by the MIT Center for Advanced Engineering, and was renamed Out of the Crisis in 1986. Deming created a system of 14 key principles for management to follow for significantly improving the effectiveness of a business or organisation. Some principles are philosophical and cultural and some are more practical. The points were first presented in his book "Out of the Crisis."

However, the realisation of lean as a field of engineering is attributed to Taiichi Ohno who established the Toyota Production System. After some years of internal use and development Toyota began to see the powerful benefits of a lean culture and lean principles. It was opened up to the wider industrial community in 1973.

Taichi Ohno's two basic principles about lean & management thinking include:

(1) The workplace needs to be transparent. Normally we are not alone in a company. Information needs to be readily available. Hoarding information by individuals is very harmful to the company.

(2) The second message requires that you start with your own workplace first, before we can discuss others.

Understanding Flow

Flow and how it relates to a manufacturing environment is an essential part of lean. Flow is not just limited to the customer at the end of the supply chain. Issues within the manufacturing system may result in supply issues and delays to the customer. The term "flow" has a broader implication and needs to be applied across the different steps within a lean manufacturing process.

For product manufacturing, flow begins with the supply and introduction of materials and components to the production line or system. The right flow levels at the right times ensure each step of the process operates efficiently and therefore the end result will be greater overall efficiencies that benefit the customer and end user. Flow is when there is no queuing or delays between each value added step. Not getting the flow levels correct can have a knock on effect right across a value stream.

For instance, (1) if queues form between value added steps, this indicates a potential bottleneck or pinch point. This may therefore call for greater capacity, e.g. more machines, faster cycle times, more operators and so on. (2) If flow is not optimised correctly, there can be an impact on the levels of WIP, intermediate product and inventory levels. High levels of inventory is costly and ties up the cash flow of a company or organisation. In order to achieve the right balance, steps that actually create value need to occur in rapid sequence. Each step should be value added. The customer does not want to pay for non-value added activities and therefore they must be eliminated or continuously reduced.

The Goal of Lean?

The ultimate goal of lean manufacturing is to achieve a perfect process. The ideal process is simply one with the perfect performance where all sources and causes of waste are reduced to zero. As previously stated, the customer is the authority on value. They determine if the right combination of quality is provided at the right place, right time and at a cost effective price. Any steps in the process should be designed to add value to the customer; these steps can range from design activities, manufacturing, packaging and so on.

Within a lean manufacturing organisation the tenets and philosophy of lean must be practised on a regular basis. Not only this, it is the responsibility of every person to embrace lean culture and practices. Remember, lean fosters continual improvement of processes.

Lean & Six Sigma

Six Sigma helps companies in identifying and controlling variation in processes that most affect performance and process outcomes. Variation in its most severe form leads to defects and defects cost money. Controlling the variation can lead to more efficiently run processes. Six sigma practitioners such as Black Belts analyse root causes and implement corrective actions A black belt project can take from 4 to 6 months to complete, however, the cost saving and return to the company can be in the hundreds of thousands of pounds, dollars or euro in value.

Toyota Production System (TPS)

Providing value to customers is a core aim of lean manufacturing. In the ideal world only value added activities are required during the manufacturing process. However, no system is likely to be perfect. When creating a value stream for a particular process or product you identify all the steps that occur to get the product or service to your customers along with key information on the steps and activities e.g. machine type, cycle times etc.

Value Stream Mapping

Value stream mapping is used to visualise and capture specific activities of a manufacturing process. It is often underestimated but listing the sequence of events or actions can raise a lot of questions within a group. Some may disagree with the sequence of steps, some may omit steps etc. However, these building blocks of the process must be noted down on paper. Above all, the best approach is to map the value stream in its current state to analyse how does it work "today", then later on the team can propose the ideal state. In simple terms, value is the worth placed upon something, either a product, service or something that a customer can express in terms of money. A key principle of lean is listening to the voice of the customer (VOC) and creating a clear picture of what value is from the customer's perspective. The customer is the one who can define the true value of a product or process and whether it is value added. However, there are 3 common principles that should be followed (1) the customer must be willing to pay for the activity (2) the activity must transform the product or service in some way and (3) the activity must be done the first time correctly. The above principles of value added activities apply to the whole process, consisting of all activities regarding people, processes systems and equipment. Applying the 3 principles consistently will allow non value added activities to be spotted quite easily. In contrast to value added activities, it is also important to understand what constitutes non value added activities. In manufacturing if an activity does not satisfy the above criteria we can determine the action to be non-value added. Simply put, this means the customer will not be willing to pay for it or the steps do not in any way transform or improve a product service or ultimately, it cannot be done correctly the first time and therefore is a waste of time and resources (which is paramount to increasing costs).

Poka Yoke

Poka Yoke is a technique for avoiding simple human error in the workplace. Simply put, it aims eliminate mistakes and is often referred to as mistake-proofing or fail-safe work methods. Poka Yoke is a system designed to prevent inadvertent errors made by workers performing a process. The word "Poka-Yoke" is Japanese for mistake-proofing or mistake avoidance. It involves the design of products, work practices, fixtures and jigs etc. that prevent the mistakes or errors that result in defects. A secondary aim of Poka Yoke is to make any defect easy to recognise with minimum time, skill and expertise. It is accepted as a simple and inexpensive way of preventing defects from being made or identifying a defect so that it is not passed to the next operation, downstream process and ultimately, the consumer.

The benefits of Poka Yoke are extensive. The specific benefits depend on the nature of the work and also where the focus is placed when executing a Poka-Yoke programme. If it focuses on cost reduction, the metric might be a decrease in set up times or processing times. If the focus is on quality, a redesign of jigs or fixture may be required. Here are 12 benefits of Poka Yoke:

- Reduces set-up issues
- Improves product quality
- Improves yield
- Reduces rework
- Reduces manufacturing cost
- Decreases set-up time
- Decreases set-up complexity
- Improves housekeeping
- Removes dependence on high skill levels or experience
- Increases manufacturing flexibility
- Improves work attitudes
- Reduces manufacturing costs

Key Principles of Poka Yoke

Workers Have Ability and Intelligence

For the application of any methodologies such as Poka Yoke, the right approach and attitude largely determines the success. The persons responsible for the rollout of such methodologies must fully support the programme and truly believe in the benefits. Typically, Top management are responsible for the rollout of programmes such as Poka-Yoke and Lean etc. Human intelligence can quickly see through any new initiatives that are not understood or supported by management.

There should also be an acknowledgement of worker's ability and aptitude. No matter what the role of the person, be it factory operator, technician, electrician or engineer, trust and confidence must be given to all stakeholders. It is a dangerous trait to overlook the input and contribution of junior staff.

Prevention Is Better Than Detection

Preventing defects saves money. Detecting defects costs money. Therefore, the preference is to prevent defects before they happen. This is where mistakes can be eliminated by Poka-Yoke techniques. If you eliminate mistakes or errors, this in turn works to eliminate defects. While inspection and detection systems will always have some use, the absence of defects will reduce the dependency of detection.

Even 1 Defect Is NOT Acceptable

Having zero defects is the golden rule. If defects are accepted as part of a culture, then complacency follows.
Remove Duplicate or Unnecessary Tasks

It may seem like a trivial observation but allowing duplicate tasks that are unnecessary (not required) can lead to mistakes causing defects. Why you may ask? Apart from the cost of completing extra manufacturing steps that are not required, extra steps or tasks generally result in extra handling. Extra handling may involve transfer to and from work stations, labelling, paperwork completion and stand down times. All of these activities can be a source of error generation.

The above figure shows the 3 elements of the triple constraint theory often used in project management. This theory is also applicable to manufacturing and mistake proofing.

Cost

Cost is one of the most important factors to today's consumer. When faced with an abundance of choice, the sale price needs to be in keeping with other manufacturers. Even if the product is of superior quality and function, cost needs to be understood and managed effectively.

With regards to Poka Yoke and mistake proofing, there is an opportunity to reduce costs during manufacturing. If errors and mistakes are prevented, this reduces defects. Not having to rework product or dump defective products helps to save money and frees up cash within a factory. If you can reduce manufacturing costs by eliminating defects and wastage, this creates an opportunity to make a higher profit margin on the market to retail price.

Quality

Some defects may result in dissatisfaction for the customer. More serious defects may prevent the customer using the product at all or in a safe manner. It is in everyone's interest to manufacture a quality product that meets the user's requirements. Product quality is often a byword for safety which should always be priority to a manufacturer.
It should be understood that changes in the level of product quality may impact upon both costs and time. It can be said that this is a balancing act, however, there should be a minimum standard of quality that is met every time.

Time

"Making mistakes increases the amount of work."

The time constraint also has implications on cost and quality. Take a manual process such as hand finishing wood. If the job is rushed or time is squeezed, the quality of the finish may not be up to standard.

If the hand finishing takes too long, it impacts upon the overhead costs associated with labour and the cost to deliver the product rises. It may be prudent to reduce the time required to complete a task or job without compromising the quality of the work.

The time to complete each manufacturing step contributes to the final cost of the product. This creates a lead time for each product type. Lead times become important in scheduling the right products and right volumes in order to meet the market requirements.

In recent years, with many consumer products, people expect next day delivery. This means the manufacturer most also be responsive and flexible to the demands of the market and wholesalers.

Customer Driven Companies

Every manufacturer wants to meet the expectations of its customers in terms of quality and other factors such as delivery times and costs. A popular standard which is used by manufactured worldwide is ISO 9000 Quality Management.

The ISO 9000 is made up of the following standards:

ISO 9001:2015 - sets out the requirements of a quality management system
ISO 9000:2015 - covers the basic concepts and language
ISO 9004:2009 - focuses on how to make a quality management system more efficient and effective
ISO 19011:2011 - sets out guidance on internal and external audits of quality management systems.
(Ref: http://www.iso.org/iso/iso_9000)

A key requirement of ISO 9000 is customer focus, the requirements of clause 5.2 deals with meeting customer requirements, and also managing the feedback from customers on an on-going basis.

Customer-Driven Practices and Quality Policies

A quality policy is a concise statement that sets out a company's commitment to the customer and the commitment to delivering quality products and services. Often a quality policy will be displayed in the reception area of a company or is available to download as a document on their website.

The quality policy must be relevant to the business operations. While many themes and traits are common across different sectors, the quality policy for a service company would likely differ slightly to a company that manufactures physical products.

Example of a Quality Policy

"We practice continual improvement to achieve customer delight by providing customer-centric, cost-effective, timely and qualitative software solutions. We are committed to meeting the regulatory requirements of medical device manufacturing, and meeting our customer expectations" Can errors be eliminated?

In the journey to achieve zero mistakes and zero defects, one school of thought promotes the belief that human error can be reduced to a minimum or indeed eliminated. There are several factors that help reduce the amount of mistakes people make. Training and experience are key parts, along with the proper systems and resources e.g. tooling, instruction, work area setup to name but a few.

Are errors inevitable?

An opposing view of errors is that people always make mistakes, no matter how small or low in occurrence. Even if we accept mistakes as a part of life, we still tend to blame the people who make them. The risk of adopting this philosophy is that defects can be missed during manufacturing which can result in defective products being sold commercially.

Sources of Errors

Poka-yoke tries to prevent or eliminate human errors. There are several common types of human errors some of which include:

Communication errors: Often mistakes occur due to a lack of communication or due to someone misinterpreting instructions. Mitigation: ensure any critical communications are written or available to review if there is doubt.

Rookie errors: Sometimes we make mistakes through lack of experience. For example, a new worker does not know the operation or is just barely familiar with it. Mitigation: Skill building and work standardisation.

Compliance Errors: If no consequences are perceived, sometimes we can overlook steps or processes. Mitigation: Foster a sense of personal responsibility and the impact of small defects on the customer.

Forgetting: Humans are prone to forgetting steps or tasks especially if they are repetitive and they are working on the same processes for long periods of time. Mitigation: Provide checklists to operators and workers in order to formalise the process. Paperwork documenting critical steps will alert the operator if they forget a step.

Procedure related errors: If instructions or standard operating procedures are inadequate it may lead to errors. Mitigation: Ensure existing work instructions are accurate and reflect the proper and necessary actions for safety, quality and prevention of mistakes.

There are various types of defects. The table below lists some common defects along with some suggested sources:

Jigs and Fixtures

The terms "jig" and "fixture" are commonly used in the manufacturing industry, particularly in CNC machining and fabrication. Many machining processes require jigs and fixtures in order to achieve consistent and accurate results.

Jig

A jig is used to guide the item or component that has to be machined while a fixture holds in place of "fixes" the component to be machined or processed.

Fixture

A fixture is used to hold the component or part during the machining process. Its purpose is not to guide the part towards the machining tool. Fixtures are secured with the table surface of the mills in most of the cases. Fixtures reduce the need for other tools and facilitate more accurate machining and processes.

Common Poka-Yoke Tools

By using Poka-yoke tools, we are trying to eliminate human errors. These errors are usually oversights due to poor judgement or concentration. Poka-yoke helps prevent defects resulting from human error or mistakes. Human factors or human errors can lead to quality defects. Poka Yoke helps people including factory operators, fabricators, assembly personal and engineers to reduce defects due to errors. Some common examples of Poka Yoke tools include:
    Checklists

Checklists are a practical and efficient way of detecting errors before they impact a process or product. Some of the best checklists are designed to be completed in a relatively short period of time, however, this can be influenced by the complexity of the task at hand. The principle still stands that designing a checklist that is to the point and easily completed will deliver the best benefits. Checklists should focus on factors that if overlooked in error can lead to defects.

Error detection - visual and audio alarms

A lot of automated pieces of equipment have in-built controls that will alarm visually and audibly when a process begins to drift or go out of control. For example, a parts washer may have temperature alarms to indicate if the temperature drops or rises below the process settings. Alarms and warning systems therefore can prevent mistakes and defects before they materialise. They can also help detect when errors do occur and help to ensure the customer gets a quality product.

☐

Guide pins

Guide pins are a proven way to help force the proper set-up and assembly of parts. Typically guide pins of different sizes are used to orient and position components in the desired manner. This prevents misoperations such as drilling or machining in the wrong position. Guide pins also help to ensure components are assembled in the correct way.

As previously described, a jig is used to guide the item or component that has to be machined while a fixture holds in place of "fixes" the component to be machined or processed. Jigs are very common and if quality upfront engineering is given to jig design, it can eliminate a lot of errors and defects during manufacturing, especially at high volumes.

Kaizen

Kaizen is the Japanese word for improvement though the term is often meant to infer not only improvement but "continuous improvement". Although the event can be technical in nature, the key components of a Kaizen event involve copper-fastening employee involvement along with the right attitude and a culture that supports improvement and lean principles. In practical terms, Kaizen can be described as a highly focused "assault" on process to problem in order to realise a rapid improvement. The event itself can take anywhere from 3-5 days to complete, so there can be quite an amount of progress made during this period. While Kaizen events are focused on improvement, they do not replace any continuous improvement programs.

Before the Kaizen event begins, it is worth spending a little time framing the issue and some key factors that may be examined during Kaizen event.

What are the goals of the Kaizen event?

- o Do the goals align with the strategic goals of the organisation?
- o Is there a comprehensive understanding of the current performance of process, quality metrics etc.?
- o Is there a high likelihood of success?
- o Will the results will be highly visible?

As with any engineering project, the involvement of management and suitable team members is essential in ensuring success. The roles and responsibilities of participants should also be determined prior to the event. You will need (1) a team sponsor, someone who can support the work of the team, typically a manager or director. The team sponsor can ensure any financing is provided along with the right resources. (2) A team leader, a point person needs to co-ordinate the event and lead the team to its goals and objectives. (3) A facilitator may be used if the team is new to what constitutes a Kaizen event. The facilitator should be experienced in Kaizen and lean and can help guide and direct the team in the techniques required. (4) Team participants, depending on the scope and project charter, various skills may be required. Match the person to the task. (5) Create a team charter; this is a great way to get input from each team member and ensure everyone is one the right page. (6) Get the data prior to the event - save time and get off to a good start. It also makes a great impression with the wider team. You don't want to be waiting on information as the event is just about to start.

Kaizen Team Charter

Team charter clearly identifies rules of:
    Operation
    Objectives
    Scope
    Resources
    Authority of the team
    Deliverables
    Schedule
    Code of Conduct
     Created by leader and approved by teamKanban

Pull

Pull is a method used to control the resources by substituting only the material consumed. The customer initiates the process or pulls the product or service. During a "pull" manufacturing system, work is started in response to a signal from the customer (or an order being received). This prevents the occurrence of costly overproduction and the development of too much WIP.

Push

In a push system, the production process is started with the order. Production orders are based on the production plan. The production start does not depend on the current production.

<u>Basic considerations</u>

The introduction of Kanban systems can be difficult to implement and it is hard to know where to begin. The following points should be understood and answered prior to roll out.

- o What is the right inventory size?
- o What is the amount of inventory required? Where?
- o What is the replenishment frequency?
- o What is the batch size?
- o Which information is essential for the Kanban signals?
- o How can we manage number and location of Kanban?
- o How can we make sure that the Kanbans are simple, visual and effective?
- o How often should the Kanbans be evaluated?
- o The system requires similar process steps and cycle times
- o These considerations may help:
- o Are there bottlenecks? Where?
- o What is the minimum batch size?

- How can we optimise setup times?
- How often does the ConWIP system have to be evaluated?
- The appliance of the theory of constraints is the simplified version of a ConWIP system
- The introduction process is effected in 3 steps:
- Identify the bottleneck
- Develop a buffer of inventory in front of the drum to keep it turning
- Release orders as they are consumed by the drum signal – the rope
- In separate step the bottleneck has to be eliminated

<u>Types of Kanban</u>

Production Kanban: allows manufacturing to replenish removed material

Withdrawal Kanban: asks for transportation of materials/product

Signal Kanban: when a buffer stock is required between work stations, because the upstream process step produces batches bigger than one Kanban (such as long setup times but very short cycle times). These triangle Kanbans work as visual signals to production/ delivery

Kanban Space – a designated area that functions similar to a Kanban shelf. When the space is empty, the quantity in the shelf has reached a minimum and production is triggered

Kanbans are typically identified with the below information at a minimum:

- Name and description of the product
- Storage location for the final product

o A Kanban number to identify it from other similar Kanbans

In a Pull system, the information for the production (after order) is only delivered to one process – the "pacemaker". The "pacemaker" starts the production and determines the rate. It is the starting point that enables a continuous flow to the customer / finished goods inventory. Inventories are essentially buffers which facilitate varying customer demand and help to absorb any demand variations to the supply chain.

Inventory size depends on:

> Demand requirements
> Variation in supply
> Takt time – cycle time ratio
> Number of different products (if applicable)

Inventories are needed where replenishment time is bigger than lead time.

Cycle stock: Average demand per period x Average lead time per period

Buffer stock: 2 standard deviations of the average demand per period

Safety stock: Average scrap per period + 2 standard deviations from the average scrap

Determining the number of Kanbans required for process or manufacturing line is done using a simple calculation. As a rule of thumb, the optimum number of Kanbans should be between 3-7 in number with a minimum of 2. The below equation can be used to determine the number of Kanbans required:

Printed in Poland
by Amazon Fulfillment
Poland Sp. z o.o., Wrocław